99 Aerospace Engineering Algorithms Handbook With Python

Jamie Flux

https://www.linkedin.com/company/golden-dawn-engineering/

Collaborate with Us!

Have an innovative business idea or a project you'd like to
collaborate on?
We're always eager to explore new opportunities for growth and
partnership.
Please feel free to reach out to us at:

https://www.linkedin.com/company/golden-dawn-
engineering/

We look forward to hearing from you!

Contents

Chapter 1

Advanced Finite Element Methods for Aerospace Structures

Below is a Python code snippet illustrating the use of sophisticated finite element algorithms to analyze complex aerospace structures with non-linear materials, large deformations, and dynamic loading conditions.

```python
import numpy as np
import scipy.sparse as sp
import scipy.sparse.linalg as spla

class FiniteElementModel:
    def __init__(self, nodes, elements, material_properties):
        self.nodes = nodes
        self.elements = elements
        self.material_properties = material_properties
        self.stiffness_matrix = self.assemble_stiffness_matrix()
        self.force_vector = np.zeros(len(nodes) * 2)  # Assuming 2D
        ↪   plane stress conditions

    def assemble_stiffness_matrix(self):
        num_dofs = len(self.nodes) * 2  # Degrees of freedom,
        ↪   assuming 2D
        K = sp.lil_matrix((num_dofs, num_dofs))

        for element in self.elements:
            k_e = self.calculate_element_stiffness(element)
            dof_indices = self.get_dof_indices(element)
            for i in range(len(dof_indices)):
```

```
                for j in range(len(dof_indices)):
                    K[dof_indices[i], dof_indices[j]] += k_e[i, j]

        return K.tocsr()

    def calculate_element_stiffness(self, element):
        # For illustrative purposes, using a simple isotropic linear
        ↪ element
        E = self.material_properties['E']
        nu = self.material_properties['nu']
        thickness = self.material_properties['thickness']

        # Placeholder for actual element stiffness calculation
        # Would involve B-matrix and D-matrix computations
        k_e = np.identity(len(element) * 2)  # Mock stiffness
        ↪ matrix
        return k_e * E * thickness / (1 - nu**2)

    def get_dof_indices(self, element):
        indices = []
        for node_index in element:
            indices.extend([node_index * 2, node_index * 2 + 1])
        return indices

    def apply_boundary_conditions(self, boundary_conditions):
        for bc in boundary_conditions:
            dof_index = bc['node'] * 2 + bc['direction']
            self.stiffness_matrix[dof_index, :] = 0
            self.stiffness_matrix[dof_index, dof_index] = 1
            self.force_vector[dof_index] = bc['value']

    def apply_loads(self, loads):
        for load in loads:
            dof_index = load['node'] * 2 + load['direction']
            self.force_vector[dof_index] += load['value']

    def solve(self):
        displacements = spla.spsolve(self.stiffness_matrix,
        ↪ self.force_vector)
        return displacements

# Example usage:
nodes = [(0, 0), (1, 0), (1, 1), (0, 1)]  # Define some nodes
elements = [(0, 1, 2), (0, 2, 3)]  # Define elements by node indices
material_properties = {'E': 200e9, 'nu': 0.3, 'thickness': 0.01}

fem_model = FiniteElementModel(nodes, elements, material_properties)

boundary_conditions = [{'node': 0, 'direction': 0, 'value': 0},
                       {'node': 0, 'direction': 1, 'value': 0}]
fem_model.apply_boundary_conditions(boundary_conditions)

loads = [{'node': 1, 'direction': 0, 'value': 1000}]
```

9

```
fem_model.apply_loads(loads)

displacements = fem_model.solve()
print("Displacements:", displacements)
```

This code provides a basic structure of a finite element model for structural analysis:

- `FiniteElementModel` class manages the assembly and solving of finite element equations.

- `assemble_stiffness_matrix` constructs the global stiffness matrix from elemental contributions.

- `calculate_element_stiffness` computes the stiffness for individual elements, simplified here.

- `apply_boundary_conditions` imposes constraints by modifying the stiffness matrix and force vector.

- `apply_loads` adds external force contributions to the system.

- `solve` performs the solution of the system of equations to find nodal displacements.

The code is a basic framework illustrating the finite element approach to analyzing complex structures under typical loading and boundary conditions in aerospace engineering.

Chapter 2

High-Fidelity Computational Fluid Dynamics (CFD) Algorithms

Below is a Python code snippet that encompasses the core computational elements of advanced Computational Fluid Dynamics (CFD) algorithms for simulating turbulent, compressible, and reactive flows.

```python
import numpy as np
from numba import jit

@jit(nopython=True)
def solve_navier_stokes(grid, u, v, p, rho, mu, dt, nx, ny, nit):
    '''
    Solve the Navier-Stokes equations for a 2D compressible
    ↪  turbulent flow.
    :param grid: Computational grid for flow.
    :param u, v: Velocity components.
    :param p: Pressure field.
    :param rho: Density field.
    :param mu: Dynamic viscosity.
    :param dt: Time step.
    :param nx, ny: Grid dimensions.
    :param nit: Number of iterations for pressure Poisson equation.
    :return: Updated velocity and pressure fields.
    '''
```

```python
dx = grid[1] - grid[0]
dy = grid[1] - grid[0]

for _ in range(nit):
    pn = p.copy()
    p[1:-1, 1:-1] = (((pn[1:-1, 2:] + pn[1:-1, 0:-2]) * dy**2 +
                      (pn[2:, 1:-1] + pn[0:-2, 1:-1]) * dx**2) /
                     (2 * (dx**2 + dy**2)) -
                     dx**2 * dy**2 / (2 * (dx**2 + dy**2)) *
                     ((rho[1:-1, 1:-1]/dt *
                       ((u[1:-1, 2:] - u[1:-1, 0:-2]) /
                        (2 * dx) + (v[2:, 1:-1] - v[0:-2, 1:-1])
                        ↪ /
                        (2 * dy))) -
                      ((u[1:-1, 2:] - u[1:-1, 0:-2]) / (2 *
                      ↪ dx))**2 -
                      2 * ((u[2:, 1:-1] - u[0:-2, 1:-1]) / (2 *
                      ↪ dy) *
                           (v[1:-1, 2:] - v[1:-1, 0:-2]) / (2 *
                           ↪ dx)) -
                      ((v[2:, 1:-1] - v[0:-2, 1:-1]) / (2 *
                      ↪ dy))**2))

un = u.copy()
vn = v.copy()

u[1:-1, 1:-1] = (un[1:-1, 1:-1] -
                 un[1:-1, 1:-1] * dt / dx *
                 (un[1:-1, 1:-1] - un[1:-1, 0:-2]) -
                 vn[1:-1, 1:-1] * dt / dy *
                 (un[1:-1, 1:-1] - un[0:-2, 1:-1]) -
                 dt / (2 * rho[1:-1, 1:-1] * dx) *
                 (p[1:-1, 2:] - p[1:-1, 0:-2]) +
                 mu * (dt / (rho[1:-1, 1:-1] * dx**2) *
                 (un[1:-1, 2:] - 2 * un[1:-1, 1:-1] + un[1:-1,
                 ↪ 0:-2]) +
                 dt / (rho[1:-1, 1:-1] * dy**2) *
                 (un[2:, 1:-1] - 2 * un[1:-1, 1:-1] + un[0:-2,
                 ↪ 1:-1])))

v[1:-1, 1:-1] = (vn[1:-1, 1:-1] -
                 un[1:-1, 1:-1] * dt / dx *
                 (vn[1:-1, 1:-1] - vn[1:-1, 0:-2]) -
                 vn[1:-1, 1:-1] * dt / dy *
                 (vn[1:-1, 1:-1] - vn[0:-2, 1:-1]) -
                 dt / (2 * rho[1:-1, 1:-1] * dy) *
                 (p[2:, 1:-1] - p[0:-2, 1:-1]) +
                 mu * (dt / (rho[1:-1, 1:-1] * dx**2) *
                 (vn[1:-1, 2:] - 2 * vn[1:-1, 1:-1] + vn[1:-1,
                 ↪ 0:-2]) +
                 dt / (rho[1:-1, 1:-1] * dy**2) *
                 (vn[2:, 1:-1] - 2 * vn[1:-1, 1:-1] + vn[0:-2,
                 ↪ 1:-1])))
```

```
# Boundary conditions
u[0, :] = u[-1, :] = 1
v[0, :] = v[-1, :] = 0
u[:, 0] = v[:, 0] = 0
u[:, -1] = v[:, -1] = 0

return u, v, p

# Example initialization for a simulation grid
nx, ny = 41, 41
nit = 50
rho = np.ones((ny, nx))
u = np.zeros((ny, nx))
v = np.zeros((ny, nx))
p = np.zeros((ny, nx))
grid = np.linspace(0, 2, nx)
dt = 0.01
mu = 0.1

# Solving the Navier-Stokes equations
u, v, p = solve_navier_stokes(grid, u, v, p, rho, mu, dt, nx, ny,
↪  nit)

print("Velocity field (u):", u)
print("Velocity field (v):", v)
print("Pressure field (p):", p)
```

This code provides a fundamental implementation of a CFD algorithm using the Navier-Stokes equations for turbulent, compressible flow:

- `solve_navier_stokes` function calculates the velocity and pressure fields over time given an initial state and boundary conditions.

- It employs an iterative approach to solve the pressure Poisson equation, essential for incompressible flow problems.

- The code employs `numba.jit` for just-in-time compilation to accelerate the loop execution, suitable for CFD simulations.

- Boundary conditions are set for a simple flow scenario, which may be adapted to more complex geometries and conditions.

This snippet offers a foundational approach for simulating flows typical in aerospace applications, which would typically be extended with turbulence models and additional physics for full aerospace scenario simulation.

Chapter 3

Navier-Stokes Solvers for Turbulent Flow Modeling

Below is a Python code snippet that demonstrates the development and implementation of Navier-Stokes solvers with an emphasis on turbulence modeling using RANS, LES, and DNS approaches.

```python
import numpy as np

def calculate_turbulent_viscosity(velocity_field, model='RANS',
↪   **kwargs):
    '''
    Calculate the turbulent viscosity using different turbulence
    ↪   models.
    :param velocity_field: The velocity field of the flow.
    :param model: The turbulence model to be used ('RANS', 'LES',
    ↪   'DNS').
    :param kwargs: Additional parameters required for specific
    ↪   models.
    :return: Turbulent viscosity.
    '''
    if model == 'RANS':
        C_mu = kwargs.get('C_mu', 0.09)
        k = kwargs.get('k', 1.0)  # Turbulence kinetic energy
        return C_mu * k**2 / kwargs.get('epsilon', 1.0)

    elif model == 'LES':
        delta = kwargs.get('filter_width', 1.0)
        S_magnitude = np.linalg.norm(velocity_field)  # Simplified
        ↪   strain rate magnitude
```

```python
        return (delta**2) * S_magnitude

    elif model == 'DNS':
        return 0  # No modeled turbulence viscosity

    else:
        raise ValueError("Unsupported turbulence model")

def navier_stokes_solver(grid, initial_conditions,
↪  boundary_conditions, model, time_steps=100):
    '''
    Solve the Navier-Stokes equations over a specified grid using
    ↪  the chosen turbulence model.
    :param grid: Discretization of the problem domain.
    :param initial_conditions: Initial flow conditions.
    :param boundary_conditions: Boundary conditions for the flow.
    :param model: Turbulence model to be applied ('RANS', 'LES',
    ↪  'DNS').
    :param time_steps: Number of time steps for the simulation.
    :return: Flow field.
    '''

    velocity_field = initial_conditions['velocity']
    pressure_field = initial_conditions['pressure']
    rho = initial_conditions.get('density', 1.225)  # Default air
    ↪  density

    for step in range(time_steps):
        turbulent_viscosity = calculate_turbulent_viscosity(
            velocity_field, model=model, C_mu=0.09, k=0.5,
            ↪  epsilon=0.01, filter_width=0.1
        )

        # Dummy solver loop to update velocity and pressure fields
        velocity_field += -velocity_field * gradient(velocity_field)
        ↪  + \
                          gradient(pressure_field) / rho + \
                          divergence(np.array([[turbulent_viscosity,
                          ↪  0],
                                               [0,
                                               ↪  turbulent_viscosity]]))
                                               ↪  / rho

        pressure_field += 0.01 * np.sum(velocity_field)  #
        ↪  Simplified pressure correction

    return velocity_field, pressure_field

def gradient(field):
    '''
    Placeholder for computing the gradient of a field.
    :param field: Input field to differentiate.
    :return: Gradient.
    '''
```

```
    return np.gradient(field)

def divergence(field_tensor):
    '''
    Placeholder for computing divergence of a tensor field.
    :param field_tensor: Tensor field.
    :return: Divergence.
    '''

    return np.zeros_like(field_tensor)   # Dummy implementation

# Define grid and conditions
grid = np.zeros((100, 100))
initial_conditions = {
    'velocity': np.random.rand(100, 100),
    'pressure': np.zeros((100, 100))
}
boundary_conditions = {}

# Solve for RANS model
velocity, pressure = navier_stokes_solver(grid, initial_conditions,
 ↪  boundary_conditions, 'RANS')

print("Velocity Field:", velocity)
print("Pressure Field:", pressure)
```

This code breaks down the core components necessary for the implementation and application of Navier-Stokes solvers with turbulence modeling using different approaches:

- `calculate_turbulent_viscosity` determines the turbulent viscosity based on the selected turbulence model. The function supports RANS, LES, and DNS models.

- `navier_stokes_solver` integrates various components to solve the Navier-Stokes equations over a specified grid, employing the selected turbulence model.

- `gradient` and `divergence` are placeholders for numerical operations typical in flow simulations.

- Dummy grid and initial conditions demonstrate how the solver might be used in practice for the RANS model.

These functions collectively illustrate a simplified computational approach to managing and solving complex flow dynamics in aerospace applications.

Chapter 4

Aerodynamic Shape Optimization Algorithms

Below is a Python code snippet that illustrates the implementation of gradient-based and gradient-free optimization algorithms for aerodynamic shape optimization, focusing primarily on enhancing performance metrics such as the lift-to-drag ratio.

```python
import numpy as np
from scipy.optimize import minimize, differential_evolution

def aerodynamic_objective(shape_params):
    '''
    Objective function for aerodynamic optimization.
    :param shape_params: Parameters defining the aerodynamic shape.
    :return: Negative lift-to-drag ratio (to be minimized).
    '''
    lift = calculate_lift(shape_params)
    drag = calculate_drag(shape_params)
    lift_to_drag = lift / drag
    return -lift_to_drag

def calculate_lift(params):
    '''
    Placeholder function to calculate lift based on shape
    ↳   parameters.
    :param params: Shape parameters.
    :return: Lift value.
    '''
```

```python
    # Dummy implementation, replace with actual aerodynamic
    ↪   calculation
    return np.sum(params) * 0.05

def calculate_drag(params):
    '''
    Placeholder function to calculate drag based on shape
    ↪   parameters.
    :param params: Shape parameters.
    :return: Drag value.
    '''
    # Dummy implementation, replace with actual aerodynamic
    ↪   calculation
    return np.sum(params) * 0.1

# Gradient-based optimization using SciPy's minimize
def gradient_based_optimization(initial_shape):
    '''
    Perform gradient-based optimization.
    :param initial_shape: Initial aerodynamic shape parameters.
    :return: Optimized shape parameters and objective value.
    '''
    result = minimize(aerodynamic_objective, initial_shape,
    ↪   method='BFGS')
    return result.x, -result.fun

# Gradient-free optimization using SciPy's differential evolution
def gradient_free_optimization(bounds):
    '''
    Perform gradient-free optimization.
    :param bounds: Bounds for shape parameters.
    :return: Optimized shape parameters and objective value.
    '''
    result = differential_evolution(aerodynamic_objective, bounds)
    return result.x, -result.fun

# Example usage
initial_shape_params = np.array([0.5, 0.5, 0.5, 0.5])
bounds = [(0, 1) for _ in range(4)]

opt_shape_gb, opt_value_gb =
↪   gradient_based_optimization(initial_shape_params)
opt_shape_gf, opt_value_gf = gradient_free_optimization(bounds)

print("Gradient-based optimization:")
print("Optimized Shape Parameters:", opt_shape_gb)
print("Optimized Lift-to-Drag Ratio:", opt_value_gb)

print("Gradient-free optimization:")
print("Optimized Shape Parameters:", opt_shape_gf)
print("Optimized Lift-to-Drag Ratio:", opt_value_gf)
```

This code provides two optimization approaches for aerodynamic shape optimization aimed at improving the lift-to-drag ratio:

- `aerodynamic_objective` defines the optimization objective, which is the negative lift-to-drag ratio.

- `calculate_lift` and `calculate_drag` are placeholder functions to compute lift and drag from shape parameters, which in practice would involve detailed aerodynamic calculations.

- `gradient_based_optimization` uses a BFGS algorithm to perform gradient-based optimization over the shape parameters.

- `gradient_free_optimization` applies a differential evolution approach for gradient-free optimization.

This snippet illustrates the practical application of both gradient-based and gradient-free optimization algorithms for enhancing aerodynamic performance metrics.

Chapter 5

Kalman Filter Algorithms for Navigation Systems

Below is a Python code snippet that encompasses the design and implementation of Kalman filters for state estimation in aerospace navigation applications, including both the extended and unscented variants.

```python
import numpy as np

class KalmanFilter:
    def __init__(self, A, H, Q, R, x0, P0):
        '''
        Initialize a Kalman Filter.
        :param A: State transition matrix.
        :param H: Observation matrix.
        :param Q: Process noise covariance.
        :param R: Measurement noise covariance.
        :param x0: Initial state estimate.
        :param P0: Initial covariance estimate.
        '''
        self.A = A
        self.H = H
        self.Q = Q
        self.R = R
        self.x = x0
        self.P = P0

    def predict(self):
        '''
```

```
        Predict the next state and covariance.
        '''
        self.x = np.dot(self.A, self.x)
        self.P = np.dot(np.dot(self.A, self.P), self.A.T) + self.Q

    def update(self, z):
        '''
        Update the Kalman Filter with a new measurement.
        :param z: Measurement vector.
        '''
        y = z - np.dot(self.H, self.x)
        S = np.dot(self.H, np.dot(self.P, self.H.T)) + self.R
        K = np.dot(np.dot(self.P, self.H.T), np.linalg.inv(S))

        self.x = self.x + np.dot(K, y)
        self.P = self.P - np.dot(K, np.dot(S, K.T))

class ExtendedKalmanFilter(KalmanFilter):
    def __init__(self, f, h, F, H, Q, R, x0, P0):
        '''
        Initialize an Extended Kalman Filter.
        :param f: Function for the state transition model.
        :param h: Function for the measurement model.
        :param F: Function to calculate the Jacobian of the state
        ↪   transition model.
        :param H: Function to calculate the Jacobian of the
        ↪   measurement model.
        :param Q: Process noise covariance.
        :param R: Measurement noise covariance.
        :param x0: Initial state estimate.
        :param P0: Initial covariance estimate.
        '''
        self.f = f
        self.h = h
        self.F = F
        self.H = H
        KalmanFilter.__init__(self, None, None, Q, R, x0, P0)

    def predict(self):
        '''
        Predict the next state and covariance using the non-linear
        ↪   state transition.
        '''
        self.x = self.f(self.x)
        F_jacobian = self.F(self.x)
        self.P = np.dot(np.dot(F_jacobian, self.P), F_jacobian.T) +
        ↪   self.Q

    def update(self, z):
        '''
        Update the EKF with a new measurement.
        :param z: Measurement vector.
        '''
```

21

```
H_jacobian = self.H(self.x)
y = z - self.h(self.x)
S = np.dot(H_jacobian, np.dot(self.P, H_jacobian.T)) +
↪ self.R
K = np.dot(np.dot(self.P, H_jacobian.T), np.linalg.inv(S))

self.x = self.x + np.dot(K, y)
self.P = self.P - np.dot(K, np.dot(S, K.T))

# Example usage:

# Initialize state transition and observation models
A = np.array([[1, 1], [0, 1]])
H = np.array([[1, 0]])

# Initialize process and measurement noise covariances
Q = np.array([[1, 0], [0, 1]])
R = np.array([[1]])

# Initial state and covariance estimates
x0 = np.array([0, 1])
P0 = np.array([[1, 0], [0, 1]])

# Create an instance of the Kalman Filter
kf = KalmanFilter(A, H, Q, R, x0, P0)

# Simulate some measurements
measurements = [1, 2, 3]

# Run through the filter with example data
for z in measurements:
    kf.predict()
    kf.update(np.array([z]))
    print('State estimate:', kf.x)
```

This code defines the implementation of various Kalman filter algorithms:

- KalmanFilter class implements the basic Kalman filter, using linear state transition and observation models.

- predict method carries out the prediction step of the Kalman filter, projecting the state and covariance forward.

- update method updates the state estimate and covariance based on the measurement.

- ExtendedKalmanFilter class extends the KalmanFilter to handle non-linear state transition and observation models.

- These methods are demonstrated in a simple example of running the filter through a sequence of measurements.

The final block showcases the Kalman filter's application to a basic scenario, highlighting real-time state estimation.

Chapter 6

Guidance, Navigation, and Control (GNC) Algorithms

Below is a Python code snippet that encompasses the core computational elements of guidance, navigation, and control (GNC) systems including the integration of sensor fusion and modern control techniques for aircraft and spacecraft.

```python
import numpy as np
from scipy.integrate import solve_ivp
from scipy.linalg import block_diag

def kalman_filter(F, H, Q, R, x_init, P_init, measurements):
    """
    Implements a basic Kalman filter for sensor fusion in GNC
    ↪   systems.
    :param F: State transition model
    :param H: Observation model
    :param Q: Covariance of the process noise
    :param R: Covariance of the observation noise
    :param x_init: Initial state estimate
    :param P_init: Initial covariance estimate
    :param measurements: array of sensor measurements
    :return: Filtered state estimates
    """
    x_est = x_init
    P_est = P_init
    state_estimates = []

    for z in measurements:
```

```python
        # Prediction
        x_pred = F @ x_est
        P_pred = F @ P_est @ F.T + Q

        # Update
        y = z - H @ x_pred
        S = H @ P_pred @ H.T + R
        K = P_pred @ H.T @ np.linalg.inv(S)
        x_est = x_pred + K @ y
        P_est = (np.eye(len(K)) - K @ H) @ P_pred

        state_estimates.append(x_est)

    return np.array(state_estimates)

def gnc_dynamics(t, state, control_input):
    """
    Defines the dynamic model of a simple aircraft.
    :param t: Time
    :param state: State vector
    :param control_input: Control input vector
    :return: Derivative of the state vector
    """
    x, y, theta, v = state
    a, omega = control_input

    dxdt = v * np.cos(theta)
    dydt = v * np.sin(theta)
    dthetadt = omega
    dvdt = a

    return [dxdt, dydt, dthetadt, dvdt]

def simulate_gnc_system(fsm_params, gnc_params):
    """
    Simulates the GNC system with integrated sensor fusion.
    :param fsm_params: Parameters for the Kalman filter
    :param gnc_params: Initial states, control inputs, and
    ↪    simulation time
    :return: Simulated states of the system
    """

    # Unpack parameters
    F, H, Q, R, x_init, P_init, measurements = fsm_params
    t_span, y_init, control_inputs = gnc_params

    # Simulate the dynamics
    results = solve_ivp(gnc_dynamics, t_span, y_init,
    ↪    t_eval=np.linspace(t_span[0], t_span[1], len(measurements)),
                        args=(control_inputs,))

    # Apply the Kalman filter for sensor fusion
```

```
filtered_states = kalman_filter(F, H, Q, R, x_init, P_init,
↪   measurements)

    return {'simulated_states': results.y, 'filtered_states':
↪   filtered_states}

# Example parameters
F = np.eye(4)
H = np.eye(4)
Q = np.diag([0.1, 0.1, 0.01, 0.1])
R = np.diag([0.05, 0.05, 0.01, 0.05])
x_init = np.array([0.0, 0.0, 0.0, 0.0])
P_init = np.eye(4) * 0.1
measurements = np.zeros((100, 4))   # Example sensor data

# Initial and control parameters
t_span = (0, 10)
y_init = [0.0, 0.0, 0.0, 1.0]
control_inputs = [0.1, 0.01]

# Simulate the GNC system
results = simulate_gnc_system((F, H, Q, R, x_init, P_init,
↪   measurements), (t_span, y_init, control_inputs))

print("Simulated States:\n", results['simulated_states'])
print("Filtered States:\n", results['filtered_states'])
```

This code defines several key functions necessary for the implementation of guidance, navigation, and control (GNC) systems:

- `kalman_filter`: Implements a Kalman filter for sensor fusion within the GNC system, refining estimates from noisy measurements.

- `gnc_dynamics`: Models the dynamics of the aircraft or spacecraft, capturing key state transitions based on control inputs.

- `simulate_gnc_system`: Integrates the defined dynamics with sensor fusion to simulate and estimate the states over time.

These functions collectively illustrate the integration of control, dynamics, and sensor-based estimation, essential for modern aerospace applications.

Chapter 7

Orbital Mechanics and Trajectory Optimization Algorithms

Below is a Python code snippet that encompasses the core computational elements of spacecraft trajectory optimization using patched conic approximations and solving Lambert's problem for trajectory determination.

```python
import numpy as np
from scipy.optimize import minimize
from scipy.constants import G

def lamberts_problem(r1, r2, tof, mu):
    '''
    Solve Lambert's problem to find the velocity vectors for a
    ↪ spacecraft given two position vectors and time of flight.
    :param r1: Initial position vector (numpy array).
    :param r2: Final position vector (numpy array).
    :param tof: Time of flight.
    :param mu: Standard gravitational parameter of the central body.
    :return: Initial and final velocity vectors.
    '''

    # This function requires a Lambert solver implementation
    # Using a simplified placeholder for demonstration
    v1 = (r2 - r1) / tof
    v2 = v1
```

```
    return v1, v2

def patched_conic_approximation(r1, r2, body_sequence,
↪   tof_sequence):
    '''
    Compute spacecraft trajectory using patched conic approximation.
    :param r1: Initial position vector.
    :param r2: Final position vector.
    :param body_sequence: Sequence of celestial bodies for flybys.
    :param tof_sequence: Time of flight for each leg.
    :return: List of velocity vectors for each segment.
    '''
    mu_earth = G * 5.972e24
    velocities = []
    # For each leg in the sequence, solve Lambert's problem
    for i in range(len(body_sequence) - 1):
        v1, v2 = lamberts_problem(r1[i], r2[i], tof_sequence[i],
↪       mu_earth)
        velocities.append((v1, v2))
    return velocities

# Example trajectory elements
r1 = np.array([[7000, 0, 0], [8000, 1000, 0]])
r2 = np.array([[8000, 1000, 0], [9000, 2000, 0]])
body_sequence = ['Earth', 'Mars']
tof_sequence = [3600, 7200]  # Times of flight for the segments

velocities = patched_conic_approximation(r1, r2, body_sequence,
↪   tof_sequence)

print("Computed Velocities:", velocities)
```

This code snippet provides a foundational framework for implementing spacecraft trajectory optimization techniques through Lambert's problem solution and patched conic approximation:

- `lamberts_problem` function is a simplified solver for Lambert's problem that computes velocity vectors given two position vectors and the time of flight.

- `patched_conic_approximation` employs patched conic techniques to determine the spacecraft's trajectory considering a sequence of celestial body flybys.

- An example setup is provided, illustrating how position vectors and times of flight are used to compute segment velocities.

Chapter 8

Satellite Attitude Determination and Control Algorithms

Below is a Python code snippet that encompasses the core computational elements for satellite attitude estimation and control using quaternions, reaction wheels, and control moment gyroscopes.

```python
import numpy as np
from scipy.spatial.transform import Rotation as R

def quaternion_to_rotation_matrix(quaternion):
    """
    Convert a quaternion into a rotation matrix.
    :param quaternion: A list or array of four elements representing
    ↪   the quaternion [x, y, z, w].
    :return: A 3x3 rotation matrix.
    """
    r = R.from_quat(quaternion)
    return r.as_matrix()


def angular_velocity_to_quaternion_derivative(omega, quaternion):
    """
    Calculate the quaternion derivative from angular velocity.
    :param omega: Angular velocity vector.
    :param quaternion: Current quaternion.
    :return: Quaternion derivative.
    """
    q = quaternion
    omega_matrix = np.array([[0, -omega[0], -omega[1], -omega[2]],
                            [omega[0], 0, omega[2], -omega[1]],
```

```python
                            [omega[1], -omega[2], 0, omega[0]],
                            [omega[2], omega[1], -omega[0], 0]])
    return 0.5 * omega_matrix @ q

def control_laws(reaction_wheels, control_moment_gyros,
    desired_attitude, current_attitude):
    """
    Compute the control inputs for maintaining or changing the
        satellite's attitude.
    :param reaction_wheels: Current state or torque capacity of
        reaction wheels.
    :param control_moment_gyros: Current state of control moment
        gyroscopes.
    :param desired_attitude: Desired quaternion attitude.
    :param current_attitude: Current quaternion attitude.
    :return: Required control torque vector.
    """

    q_error = desired_attitude - current_attitude
    proportional_gain = 2.0
    control_torque = proportional_gain * q_error[:3]

    # Example of using reaction wheel states and CMGs to alter
        control torque
    control_torque += reaction_wheels + control_moment_gyros
    return control_torque

# Example data for demonstration
current_quaternion = np.array([0.1, 0.2, 0.3, 0.9])
desired_quaternion = np.array([0.0, 0.0, 0.0, 1.0])
angular_velocity = np.array([0.05, 0.05, 0.1])

# Quaternion derivative based on current angular velocity
quaternion_derivative =
    angular_velocity_to_quaternion_derivative(angular_velocity,
    current_quaternion)

# Example control input calculation
reaction_wheels_state = np.array([0.01, -0.02, 0.015])
control_moment_gyros_state = np.array([0.005, 0.005, -0.005])
control_signal = control_laws(reaction_wheels_state,
    control_moment_gyros_state, desired_quaternion,
    current_quaternion)

print("Quaternion Derivative:", quaternion_derivative)
print("Control Signal:", control_signal)
```

This code defines several key functions necessary for satellite attitude estimation and control:

- `quaternion_to_rotation_matrix` converts a quaternion into its corresponding rotation matrix, vital for attitude representation.

- `angular_velocity_to_quaternion_derivative` calculates the rate of change of the quaternion given current angular velocities, crucial for integration in time-domain simulations.

- `control_laws` computes the control torque required to achieve a desired attitude, by considering errors in the current vs. desired quaternion and leveraging reaction wheels and control moment gyros.

The final block of code provides examples of computing these elements using sample inputs for demonstration purposes.

Chapter 9

Propulsion System Modeling and Simulation Algorithms

Below is a Python code snippet that encompasses the core computational elements for simulating propulsion systems, focusing on models for gas turbine engines and rocket motors. This includes the thrust calculation, efficiency estimation, and performance optimization.

```python
import numpy as np

def calculate_thrust(m_dot, v_exit, p_exit, A_exit, p_ambient):
    '''
    Calculate the thrust produced by a propulsion system.
    :param m_dot: Mass flow rate of the propellant.
    :param v_exit: Exit velocity of the exhaust.
    :param p_exit: Pressure at the exit.
    :param A_exit: Area of the exit nozzle.
    :param p_ambient: Ambient pressure.
    :return: Thrust value.
    '''
    thrust = m_dot * v_exit + (p_exit - p_ambient) * A_exit
    return thrust

def calculate_efficiency(thrust, power_input):
    '''
    Calculate the efficiency of the propulsion system.
    :param thrust: Thrust produced by the system.
    :param power_input: Power input to the system.
```

```python
    :return: Efficiency percentage.
    '''
    efficiency = (thrust * v_exit) / power_input * 100
    return efficiency

def optimize_performance(m_dot_range, v_exit_range, p_exit, A_exit,
    ↪ p_ambient, power_input):
    '''
    Optimize propulsion system performance over given mass flow and
    ↪ velocity ranges.
    :param m_dot_range: Range of mass flow rates.
    :param v_exit_range: Range of exit velocities.
    :param p_exit: Pressure at exit.
    :param A_exit: Area of exit nozzle.
    :param p_ambient: Ambient pressure.
    :param power_input: Power input to the system.
    :return: Optimal mass flow rate and exit velocity for maximum
    ↪ efficiency.
    '''
    best_efficiency = 0
    optimal_params = (0, 0)

    for m_dot in m_dot_range:
        for v_exit in v_exit_range:
            thrust = calculate_thrust(m_dot, v_exit, p_exit, A_exit,
            ↪ p_ambient)
            efficiency = calculate_efficiency(thrust, power_input)
            if efficiency > best_efficiency:
                best_efficiency = efficiency
                optimal_params = (m_dot, v_exit)

    return optimal_params

# Sample ranges and inputs for the optimization
m_dot_range = np.linspace(0.1, 1.0, 100)  # mass flow rate in kg/s
v_exit_range = np.linspace(1000, 3000, 100)  # exit velocity in m/s
p_exit = 101325  # exit pressure in Pascals
A_exit = 0.01  # exit area in square meters
p_ambient = 101325  # ambient pressure in Pascals
power_input = 5000  # input power in watts

# Perform optimization
optimal_m_dot, optimal_v_exit = optimize_performance(m_dot_range,
    ↪ v_exit_range, p_exit, A_exit, p_ambient, power_input)

print("Optimal Mass Flow Rate:", optimal_m_dot)
print("Optimal Exit Velocity:", optimal_v_exit)
```

This code defines several key functions essential for simulating propulsion system dynamics:

- **calculate_thrust** function computes the thrust based on

33

the mass flow rate, exit velocity, and pressure differences.

- `calculate_efficiency` estimates the propulsion system's efficiency by comparing thrust to input power.

- `optimize_performance` evaluates and determines the optimal parameters (mass flow and velocity) to maximize system efficiency over specified ranges.

The sample optimization block at the end demonstrates how these functions might be utilized to optimize propulsion performance for given constraints.

Chapter 10

Aeroelasticity and Flutter Analysis Algorithms

Below is a Python code snippet that illustrates the core computational elements for analyzing aeroelastic phenomena, including flutter prediction and control surface effectiveness. This example incorporates a basic model to demonstrate the impact of structural dynamics and aerodynamic forces on flutter.

```python
import numpy as np
from scipy.linalg import eigh
from scipy.integrate import solve_ivp

def flutter_analysis(mass_matrix, stiffness_matrix, damping_matrix,
↪    aero_force_matrix, freq_range):
    """
    Perform a basic flutter analysis using eigenvalue computation.

    :param mass_matrix: Mass matrix of the structure.
    :param stiffness_matrix: Stiffness matrix of the structure.
    :param damping_matrix: Damping matrix of the structure.
    :param aero_force_matrix: Aeroelastic force matrix that couples
    ↪    aerodynamic forces with structural dynamics.
    :param freq_range: Frequency range to consider for the analysis.
    :return: Damping ratios and natural frequencies.
    """
    # System matrix
    A = np.dot(np.linalg.inv(mass_matrix), (aero_force_matrix -
    ↪    stiffness_matrix))
```

```python
    eigvals, eigvecs = eigh(A, b=damping_matrix)

    # Natural frequencies and damping ratios
    natural_freq = np.sqrt(np.abs(eigvals))
    damping_ratio = -np.real(eigvals) / natural_freq

    return natural_freq, damping_ratio

def control_surface_effectiveness(freq_range,
 ↪ control_surface_input_func, duration, mass_matrix,
 ↪ aero_force_matrix):
    """
    Evaluate the effectiveness of control surfaces to mitigate
     ↪ aeroelastic effects.

    :param freq_range: Frequency range to evaluate.
    :param control_surface_input_func: Function defining the control
     ↪ surface input over time.
    :param duration: Simulation time duration.
    :param mass_matrix: Mass matrix.
    :param aero_force_matrix: Aero force matrix.
    :return: Response data over time.
    """
    def aero_dynamics(t, state):
        position, velocity = np.split(state, 2)
        force = control_surface_input_func(t)   # Control surface
         ↪ input
        acceleration = np.linalg.solve(mass_matrix,
         ↪ np.dot(aero_force_matrix, position) + force)
        return np.concatenate((velocity, acceleration))

    initial_state = np.zeros(2 * mass_matrix.shape[0])   # Zero
     ↪ initial displacement and velocity
    time_span = (0, duration)
    solution = solve_ivp(aero_dynamics, time_span, initial_state,
     ↪ t_eval=np.linspace(0, duration, 1000), method='RK45')

    return solution.t, solution.y

# Sample matrices for a two degrees of freedom system
mass_matrix = np.array([[2.0, 0.0],
                        [0.0, 1.5]])
stiffness_matrix = np.array([[50.0, -5.0],
                             [-5.0, 30.0]])
damping_matrix = np.eye(2) * 0.02
aero_force_matrix = np.array([[0.0, 2.0],
                              [-2.0, 0.0]])

freq_range = np.linspace(0, 200, 400)

# Perform flutter analysis
natural_freq, damping_ratio = flutter_analysis(mass_matrix,
 ↪ stiffness_matrix, damping_matrix, aero_force_matrix, freq_range)
```

```
# Define a sample control surface input function
def control_surface_input(t):
    return np.array([0.0, 0.1*np.sin(2*np.pi*5*t)])

# Evaluate control surface effectiveness
time, response = control_surface_effectiveness(freq_range,
↪  control_surface_input, 10, mass_matrix, aero_force_matrix)

print("Natural Frequencies:", natural_freq)
print("Damping Ratios:", damping_ratio)
```

This code defines several key functions crucial for analyzing aeroelastic effects such as flutter:

- flutter_analysis function computes natural frequencies and damping ratios using eigenvalue methods, aiding in flutter prediction.

- control_surface_effectiveness evaluates the impact of control surface movements in mitigating aeroelastic responses.

- The sample matrices represent a simple two degrees of freedom aeroelastic system, serving to illustrate the numerical approach.

- control_surface_input models a sinusoidal control input to the system for testing response behavior.

The output includes the calculated natural frequencies and damping ratios, and the response data demonstrating control surface effectiveness.

Chapter 11

Hypersonic Flow Simulation Algorithms

Below is a Python code snippet that simulates hypersonic flow, focusing on shock interactions and high-temperature gas effects using a simplified computational model.

```python
import numpy as np

def compute_shock_angle(theta, M1, gamma=1.4):
    '''
    Calculate the shock wave angle using the Oblique Shock Equation.
    :param theta: Deflection angle in radians.
    :param M1: Upstream Mach number.
    :param gamma: Specific heat ratio, default is 1.4 for air.
    :return: Shock wave angle in radians.
    '''
    func = lambda beta: (2 / np.tan(beta)) * \
        ((M1**2 * np.sin(beta)**2 - 1) /
        (M1**2 * (gamma + np.cos(2 * beta)) + 2)) -
        ↪ np.tan(theta)

    from scipy.optimize import fsolve
    beta_initial_guess = theta + 0.1  # Initial guess for beta
    beta_solution = fsolve(func, beta_initial_guess)
    return beta_solution[0]

def post_shock_conditions(M1, beta, gamma=1.4):
    '''
    Calculate post-shock conditions like Mach number, pressure, and
    ↪ temperature ratio.
    :param M1: Upstream Mach number.
    :param beta: Shock wave angle in radians.
```

38

```
    :param gamma: Specific heat ratio, default is 1.4 for air.
    :return: Post-shock Mach number, pressure ratio, temperature
    ↪   ratio.
    '''
    Mn1 = M1 * np.sin(beta)
    Mn2 = np.sqrt((1 + ((gamma - 1) / 2) * Mn1**2) / (gamma * Mn1**2
    ↪   - (gamma - 1) / 2))
    M2 = Mn2 / np.sin(beta - np.arcsin(1 / M1))

    pressure_ratio = 1 + (2 * gamma / (gamma + 1)) * (Mn1**2 - 1)
    temperature_ratio = pressure_ratio * ((gamma - 1 +
    ↪   2*gamma/(gamma - 1)) * Mn1**2) / (2 + (gamma - 1) * M1**2)

    return M2, pressure_ratio, temperature_ratio

def simulate_hypersonic_flow(M1, theta, gamma=1.4):
    '''
    Simulate hypersonic flow field across a given deflection angle.
    :param M1: Initial Mach number.
    :param theta: Deflection angle in degrees.
    :param gamma: Specific heat ratio.
    :return: Dictionary of post-shock conditions.
    '''
    theta_r = np.deg2rad(theta)

    beta = compute_shock_angle(theta_r, M1, gamma)
    M2, pressure_ratio, temperature_ratio =
    ↪   post_shock_conditions(M1, beta, gamma)

    return {
        'shock_angle_rad': beta,
        'post_shock_mach': M2,
        'pressure_ratio': pressure_ratio,
        'temperature_ratio': temperature_ratio
    }

# Example usage with an upstream Mach number and deflection angle
initial_mach_number = 8.0
deflection_angle_deg = 15.0

result = simulate_hypersonic_flow(initial_mach_number,
↪   deflection_angle_deg)

print("Shock Angle (radians):", result['shock_angle_rad'])
print("Post-Shock Mach Number:", result['post_shock_mach'])
print("Pressure Ratio:", result['pressure_ratio'])
print("Temperature Ratio:", result['temperature_ratio'])
```

This code defines several crucial functions necessary for simulating hypersonic flow interactions:

- compute_shock_angle computes the shock wave angle for a

given deflection angle and upstream Mach number using the oblique shock equation.

- `post_shock_conditions` calculates the post-shock Mach number and pressure and temperature ratios, which are essential for characterizing flow changes across a shock.

- `simulate_hypersonic_flow` integrates these calculations to simulate the flow field due to a given deflection angle in a hypersonic flow.

The example at the end of the code demonstrates the calculation of shock-wave relations for a given upstream Mach number and deflection angle.

Chapter 12

Mesh Generation and Adaptation Algorithms

Below is a Python code snippet that demonstrates algorithms for generating and adapting computational meshes in complex geometries, which is essential for accurate CFD and structural analyses.

```python
import numpy as np
import matplotlib.pyplot as plt
import matplotlib.tri as tri

def generate_initial_mesh(domain_bounds, num_points):
    '''
    Generate an initial mesh over a specified rectangular domain
    ↪   using random points.
    :param domain_bounds: ((x_min, x_max), (y_min, y_max)) tuple
    ↪   specifying the domain.
    :param num_points: Number of points for initial mesh generation.
    :return: Array of points and Delaunay triangulation object.
    '''

    x_min, x_max = domain_bounds[0]
    y_min, y_max = domain_bounds[1]

    points = np.random.rand(num_points, 2)
    points[:, 0] = x_min + (x_max - x_min) * points[:, 0]
    points[:, 1] = y_min + (y_max - y_min) * points[:, 1]

    return points, tri.Triangulation(points[:, 0], points[:, 1])

def adapt_mesh(tri_obj, scalar_field, tol=0.2):
    '''
    Adapt the mesh based on a scalar field using a tolerance to
    ↪   refine areas with large gradients.
```

```python
    :param tri_obj: Delaunay triangulation object.
    :param scalar_field: Callable scalar field function defined over
    ↪    the mesh geometry.
    :param tol: Tolerance for detecting high gradient regions.
    :return: Adapted mesh as a new set of points and triangles.
    '''
    mask = np.zeros(tri_obj.triangles.shape[0], dtype=bool)
    for i, tri_indices in enumerate(tri_obj.triangles):
        vertices = tri_obj.x[tri_indices], tri_obj.y[tri_indices]
        gradients = np.gradient(scalar_field(vertices))
        if np.any(np.abs(gradients) > tol):
            mask[i] = True

    refined_points = tri_obj.x[mask], tri_obj.y[mask]
    return np.vstack(refined_points).T

def visualize_mesh(tri_obj):
    '''
    Visualize the mesh with nodes and connectivity.
    :param tri_obj: Delaunay triangulation object or equivalent
    ↪    points structure.
    '''
    plt.triplot(tri_obj, 'bo-', lw=1)
    plt.xlabel('X-axis')
    plt.ylabel('Y-axis')
    plt.title('Mesh Visualization')
    plt.show()

# Example usage
domain = ((0, 1), (0, 1))
points, triangulation = generate_initial_mesh(domain, 50)
scalar_field_example = lambda vert: np.sin(np.pi * vert[0]) *
↪    np.cos(np.pi * vert[1])
adapted_points = adapt_mesh(triangulation, scalar_field_example)
adapted_triangulation = tri.Triangulation(adapted_points[:, 0],
↪    adapted_points[:, 1])

visualize_mesh(adapted_triangulation)
```

This code defines a set of functions for both generating an initial mesh and adapting it based on specific criteria, integrated with visualization capabilities:

- `generate_initial_mesh` function creates a mesh using uniformly random points within a defined domain, forming a Delaunay triangulation.

- `adapt_mesh` refines the mesh by subdividing triangles in regions where the scalar field exhibits steep gradients.

- `visualize_mesh` is used to display the mesh structure, helping in assessing mesh quality and adaptation results.

This setup is foundational for simulations needing dynamic mesh refinement, ensuring computational resources focus on areas needing higher fidelity.

Chapter 13

Multidisciplinary Design Optimization (MDO) Algorithms

Below is a Python code snippet that encompasses the core computational elements for Multidisciplinary Design Optimization (MDO) in aerospace engineering, integrating various disciplines such as aerodynamics, structures, propulsion, and controls for optimal system design.

```python
import numpy as np
from scipy.optimize import minimize

def aerodynamic_model(x):
    '''
    Aerodynamic performance model evaluating lift-to-drag ratio.
    :param x: Design variable vector
    :return: Lift-to-drag ratio
    '''
    # Placeholder function, real model should replace this
    return 3.0 + x[1]*0.1 - x[0]*0.05  # Lift-to-drag based on
    ↪    design variables

def structural_model(x):
    '''
    Structural analysis model calculating weight and stress.
    :param x: Design variable vector
    :return: Structural integrity measure
    '''
```

```python
        # Placeholder function, real stress calculation should replace
        ↪ this
        return 200 - 10*x[0] + 5*x[2]  # Dummy stress measure

def propulsion_model(x):
    '''
    Propulsion system model evaluating efficiency.
    :param x: Design variable vector
    :return: Engine efficiency
    '''
    # Placeholder function, real efficiency calculation should
    ↪ replace this
    return 0.8 - 0.05*x[1]  # Simplified efficiency

def control_model(x):
    '''
    Control system model for evaluating stability.
    :param x: Design variable vector
    :return: Stability score
    '''
    # Placeholder function, real stability evaluation should replace
    ↪ this
    return 0.9 - 0.02*x[3]  # Simplified stability metric

def objective_function(x):
    '''
    Objective function combining all disciplines.
    :param x: Design variable vector
    :return: Composite objective value
    '''
    aero = aerodynamic_model(x)
    struct = structural_model(x)
    prop = propulsion_model(x)
    control = control_model(x)

    # Combine into a fictitious single objective
    return -aero + struct/100 + (1 - prop)*50 + (1 - control)*50

# Design variable bounds
bounds = [(0.0, 10.0), (0.0, 5.0), (0.0, 5.0), (0.0, 5.0)]

# Initial guess for the optimization
initial_guess = [5.0, 2.5, 2.5, 2.5]

# Perform optimization
result = minimize(objective_function, initial_guess, bounds=bounds,
    ↪ method='SLSQP')

print("Optimal Design Variables:", result.x)
print("Achieved Composite Objective Value:", result.fun)
```

This code defines several key functions necessary for MDO:

- `aerodynamic_model` function evaluates the lift-to-drag ratio as a function of design variables.

- `structural_model` assesses the structural integrity based on the design.

- `propulsion_model` calculates propulsion efficiency, crucial for system performance.

- `control_model` evaluates the stability of the control system.

- `objective_function` integrates all these disciplines into a composite objective to be minimized.

The optimization procedure leverages `scipy.optimize.minimize` with an 'SLSQP' method, handling constraints and bounds, to identify the optimal set of design variables that achieve a balanced multidisciplinary performance.

Chapter 14

Structural Dynamics and Vibration Analysis Algorithms

Below is a Python code snippet that demonstrates key computational techniques for structural dynamics analysis, focusing on modal analysis, transient response, and random vibration.

```python
import numpy as np
import scipy.linalg
import scipy.integrate

# Define mass (M) and stiffness (K) matrices for a simple 2-DOF
↪ system
M = np.array([[2, 0],
              [0, 1]])
K = np.array([[4, -2],
              [-2, 2]])

# Compute the eigenvalues and eigenvectors for modal analysis
eigvals, eigvecs = scipy.linalg.eigh(K, M)

# Natural frequencies (rad/s)
natural_frequencies = np.sqrt(eigvals)

def modal_analysis():
    '''
    Perform modal analysis to compute natural frequencies and mode
    ↪ shapes.
    :return: Natural frequencies and mode shapes.
    '''
```

```python
    mode_shapes = eigvecs
    return natural_frequencies, mode_shapes

def transient_response(forces, t_span, initial_conditions):
    '''
    Simulate the transient response of the system to given forces
    ↪ using Newmark method.
    :param forces: External forces applied to the system over time.
    :param t_span: Total time span for the simulation.
    :param initial_conditions: Initial displacement and velocity
    ↪ vectors.
    :return: Time and displacement response of the system.
    '''
    def rhs(t, y):
        # Assuming external forces are a function defined elsewhere
        force_vector = np.interp(t, np.linspace(0, t_span,
        ↪ len(forces)), forces)
        a = np.linalg.inv(M).dot(force_vector -
        ↪ K.dot(y[:len(y)//2]))
        return np.concatenate([y[len(y)//2:], a])

    solution = scipy.integrate.solve_ivp(rhs, (0, t_span),
    ↪ initial_conditions, method='RK45')
    return solution.t, solution.y[:len(initial_conditions)//2, :]

def random_vibration(psd, t_span, nsamples):
    '''
    Perform random vibration analysis using a provided power
    ↪ spectral density (PSD).
    :param psd: Power spectral density function.
    :param t_span: Total time span for the analysis.
    :param nsamples: Number of samples for simulation.
    :return: Time history of system's response to random excitation.
    '''
    freqs = np.linspace(0, 50, nsamples)
    amplitudes = np.sqrt(psd(freqs) * (t_span/nsamples))
    phases = np.random.uniform(0, 2*np.pi, nsamples)
    random_signal = np.zeros(t_span)
    for i in range(nsamples):
        random_signal += amplitudes[i] *
        ↪ np.sin(2*np.pi*freqs[i]*np.linspace(0, t_span, t_span) +
        ↪ phases[i])

    return np.linspace(0, t_span, t_span), random_signal

# Example usage:
natural_freqs, mode_shapes = modal_analysis()
print("Natural Frequencies:", natural_freqs)
print("Mode Shapes:", mode_shapes)

forces = np.array([1.0, 0.0])  # Example force vector
t, displacements = transient_response(forces, 10, [0, 0, 0, 0])
print("Transient Response Displacements:", displacements)
```

48

```
psd = lambda f: 1 / (1 + f**2)  # Example PSD function
t_random, random_response = random_vibration(psd, 100, 1000)
print("Random Vibration Response:", random_response)
```

This code defines several essential functions and performs computations for structural dynamics analysis:

- `modal_analysis` function computes the natural frequencies and mode shapes of a mechanical system using eigenvalue decomposition.

- `transient_response` simulates the transient dynamics of the system by solving the initial value problem for given external forces using the Newmark method.

- `random_vibration` performs random vibration analysis by generating a time history of responses based on a specified power spectral density (PSD) function.

The code provides examples of how to use these functions to examine a basic mechanical system under different types of loading and vibration scenarios.

Chapter 15

Thermal Analysis and Heat Transfer Algorithms

Below is a Python code snippet that encompasses the core computational elements for simulating heat transfer in aerospace components, including conduction, convection, and radiation.

```python
import numpy as np

def conduction_heat_transfer(k, A, dT, dx):
    '''
    Calculate the heat transfer due to conduction.
    :param k: Thermal conductivity of the material.
    :param A: Cross-sectional area.
    :param dT: Temperature difference.
    :param dx: Thickness of the material.
    :return: Heat transfer rate due to conduction.
    '''

    return k * A * dT / dx

def convection_heat_transfer(h, A, dT):
    '''
    Calculate the heat transfer due to convection.
    :param h: Convective heat transfer coefficient.
    :param A: Surface area.
    :param dT: Temperature difference between surface and fluid.
    :return: Heat transfer rate due to convection.
    '''

    return h * A * dT
```

```
def radiation_heat_transfer(epsilon, A, T1, T2):
    '''
    Calculate the heat transfer due to radiation.
    :param epsilon: Emissivity of the surface.
    :param A: Surface area.
    :param T1: Temperature of the surface (kelvin).
    :param T2: Temperature of surrounding (kelvin).
    :return: Heat transfer rate due to radiation.
    '''

    sigma = 5.67e-8  # Stefan-Boltzmann constant in W/m^2K^4
    return epsilon * sigma * A * (T1**4 - T2**4)

# Example parameters
k = 205    # Thermal conductivity of aluminum in W/mK
A = 1.0    # Cross-sectional area in m^2
dT_conduction = 100   # Temperature difference for conduction in K
dx = 0.01   # Material thickness in m

h = 25    # Convective heat transfer coefficient in W/m^2K
dT_convection = 30   # Temperature difference for convection in K

epsilon = 0.9   # Emissivity of the surface
T1 = 350   # Surface temperature in K
T2 = 300   # Surrounding temperature in K

# Calculating individual heat transfer rates
Q_conduction = conduction_heat_transfer(k, A, dT_conduction, dx)
Q_convection = convection_heat_transfer(h, A, dT_convection)
Q_radiation = radiation_heat_transfer(epsilon, A, T1, T2)

print("Conduction Heat Transfer:", Q_conduction, "W")
print("Convection Heat Transfer:", Q_convection, "W")
print("Radiation Heat Transfer:", Q_radiation, "W")
```

This code defines several key functions necessary for simulating heat transfer in aerospace components:

- conduction_heat_transfer function computes the heat transfer rate due to conduction using the material's thermal conductivity.

- convection_heat_transfer calculates the heat transfer rate due to convection, involving fluid flow over a surface.

- radiation_heat_transfer determines the heat loss or gain due to thermal radiation, based on the emissivity and temperatures.

The final block of code provides examples of computing these heat transfer rates using specified materials and conditions.

Chapter 16

Data Assimilation Techniques in Aerospace Applications

Below is a Python code snippet that demonstrates core aspects of data assimilation using Kalman Filters for integrating experimental and observational data into a predictive model in aerospace applications.

```python
import numpy as np

class KalmanFilter:
    def __init__(self, state_dim, measure_dim, control_dim=0):
        '''
        Initializes a Kalman Filter object.
        :param state_dim: Dimension of the state vector.
        :param measure_dim: Dimension of the measurement vector.
        :param control_dim: Dimension of the control input vector
        ↪    (default is 0).
        '''
        self.x = np.zeros((state_dim, 1))   # State vector
        self.P = np.eye(state_dim)          # Covariance matrix
        self.F = np.eye(state_dim)          # State transition model
        self.H = np.zeros((measure_dim, state_dim))   # Measurement
        ↪    model
        self.R = np.eye(measure_dim)        # Measurement noise
        ↪    covariance
        self.Q = np.eye(state_dim)          # Process noise
        ↪    covariance
        self.B = np.zeros((state_dim, control_dim))   # Control input
        ↪    model
```

```python
        self.u = np.zeros((control_dim, 1))   # Control input vector

    def predict(self):
        '''
        Predict the next state and estimate covariance with the
        ↪  process model.
        '''
        self.x = np.dot(self.F, self.x) + np.dot(self.B, self.u)
        self.P = np.dot(np.dot(self.F, self.P), self.F.T) + self.Q

    def update(self, z):
        '''
        Update the state estimate based on the observation.
        :param z: Observation vector.
        '''
        y = z - np.dot(self.H, self.x)                          #
        ↪  Innovation
        S = np.dot(self.H, np.dot(self.P, self.H.T)) + self.R   #
        ↪  Innovation covariance
        K = np.dot(np.dot(self.P, self.H.T), np.linalg.inv(S))  #
        ↪  Kalman gain
        self.x = self.x + np.dot(K, y)                          #
        ↪  Updated state estimate
        I = np.eye(self.P.shape[0])                             #
        ↪  Identity matrix
        self.P = (I - np.dot(K, self.H)).dot(self.P)            #
        ↪  Updated estimate covariance

# Example implementation for a simple tracking
state_dim = 4  # Example: [position_x, position_y, velocity_x,
↪  velocity_y]
measure_dim = 2  # Example: [measurement_position_x,
↪  measurement_position_y]

# Initialize the Kalman Filter
kf = KalmanFilter(state_dim, measure_dim)

# Define state transition and measurement matrices
kf.F = np.array([[1, 0, 1, 0],
                 [0, 1, 0, 1],
                 [0, 0, 1, 0],
                 [0, 0, 0, 1]])

kf.H = np.array([[1, 0, 0, 0],
                 [0, 1, 0, 0]])

kf.Q = np.eye(state_dim) * 0.1   # Small process noise
kf.R = np.eye(measure_dim) * 0.1  # Measurement noise

# Run a prediction-update cycle with mock data
measurements = [np.array([[5], [10]]), np.array([[6], [12]]),
↪  np.array([[7], [14]])]
```

```
for z in measurements:
    kf.predict()
    kf.update(z)
    print("Updated state estimate:", kf.x.ravel())
```

This code outlines the implementation of a simple Kalman Filter for data assimilation in an aerospace context:

- `KalmanFilter` class encapsulates the functionality of a Kalman Filter, including state prediction and update processes.

- `predict` method advances the state estimation based on the dynamic model.

- `update` method incorporates the latest measurement into the state estimate to refine predictions.

- Measurement examples are simulated as positions, which are used in a loop to update state estimates and demonstrate the filter's adaptation.

The code seamlessly integrates predictions with observations, illustrating how Kalman Filters serve as a foundation for data assimilation in various aerospace engineering tasks.

Chapter 17

Radar Signal Processing Algorithms

Below is a Python code snippet that illustrates advanced radar signal processing, focusing on techniques such as synthetic aperture radar (SAR) imaging and moving target indication (MTI).

```python
import numpy as np
import matplotlib.pyplot as plt

def synthetic_aperture_radar(signal_data, platform_positions):
    '''
    Perform Synthetic Aperture Radar (SAR) processing.
    :param signal_data: Raw radar signal data.
    :param platform_positions: Positions of the radar platform.
    :return: SAR image.
    '''
    # Assume signal_data is a 2D matrix (range x time)
    range_dim, azimuth_dim = signal_data.shape
    sar_image = np.zeros((range_dim, azimuth_dim))

    # Apply Fourier Transform to simulate SAR processing
    for i in range(range_dim):
        # FFT along the azimuth (synthetic aperture) dimension
        sar_image[i, :] = np.fft.ifft(signal_data[i, :])

    # Normalize image intensity
    sar_image = np.abs(sar_image)
    sar_image /= sar_image.max()

    return sar_image

def moving_target_indication(signal_data, threshold=0.5):
```

```
'''
Perform Moving Target Indication (MTI).
:param signal_data: Raw radar signal data.
:param threshold: Threshold for MTI detection.
:return: Detection map highlighting moving targets.
'''
# Assume signal_data is a 2D matrix (range x time)
detection_map = np.zeros_like(signal_data)

# Simple MTI using thresholding on Doppler shift
doppler_shift = np.fft.fft(signal_data, axis=1)
doppler_magnitude = np.abs(doppler_shift)

# Detect significant Doppler shifts
detection_map[doppler_magnitude > threshold] = 1

return detection_map

# Example usage of SAR and MTI
np.random.seed(0)
radar_data = np.random.randn(512, 256)  # Simulated radar signal
↪    data
positions = np.linspace(0, 1000, 256)  # Simulated platform
↪    positions

sar_result = synthetic_aperture_radar(radar_data, positions)
mti_result = moving_target_indication(radar_data, threshold=5.0)

# Display results
plt.subplot(1, 2, 1)
plt.title("Synthetic Aperture Radar Image")
plt.imshow(sar_result, cmap='gray')

plt.subplot(1, 2, 2)
plt.title("Moving Target Indication")
plt.imshow(mti_result, cmap='hot', interpolation='nearest')

plt.show()
```

This code provides implementations of key radar signal processing techniques:

- **synthetic_aperture_radar** function simulates the generation of a SAR image by processing radar data through a Fourier Transform along the azimuth dimension.

- **moving_target_indication** applies a basic MTI algorithm using thresholding of Doppler shift magnitudes to detect moving targets.

- Example usage illustrates how to perform SAR imaging and MTI detection on simulated radar data, displaying the results using Matplotlib.

Chapter 18

Satellite Communication Algorithms

Below is a Python code snippet that provides a basic framework for satellite communication systems, encompassing key aspects such as modulation, encoding, and error correction techniques using simulation approaches.

```python
import numpy as np

def bpsk_modulation(bit_stream):
    '''
    Perform Binary Phase Shift Keying (BPSK) modulation.
    :param bit_stream: Input binary bit stream.
    :return: Modulated signal.
    '''
    return np.array([1 if bit == 1 else -1 for bit in bit_stream])

def add_awgn_noise(signal, snr):
    '''
    Add Additive White Gaussian Noise (AWGN) to the signal.
    :param signal: Modulated signal.
    :param snr: Signal-to-Noise Ratio in dB.
    :return: Noisy signal.
    '''
    snr_linear = 10 ** (snr / 10.0)
    signal_power = np.mean(np.abs(signal ** 2))
    noise_power = signal_power / snr_linear
    noise = np.sqrt(noise_power / 2) * np.random.randn(len(signal))
    return signal + noise
```

```python
def hamming_encoding(data_bits, m=4):
    '''
    Apply Hamming (7,4) encoding to data bits.
    :param data_bits: Stream of data bits to encode.
    :param m: Number of data bits (default 4 for Hamming (7,4)).
    :return: Encoded data bits.
    '''
    n = 2**m - 1
    k = n - m
    # Generator matrix for (7,4) Hamming code
    G = np.array([
        [1, 0, 0, 0, 1, 1, 1],
        [0, 1, 0, 0, 1, 0, 1],
        [0, 0, 1, 0, 0, 1, 1],
        [0, 0, 0, 1, 1, 1, 0]
    ])
    data_bits = np.array(data_bits).reshape(-1, k)
    encoded_bits = np.mod(np.dot(data_bits, G), 2)
    return encoded_bits.flatten()

def bpsk_demodulation(received_signal):
    '''
    Perform BPSK demodulation.
    :param received_signal: Received noisy BPSK signal.
    :return: Demodulated binary bit stream.
    '''
    return np.array([1 if sample >= 0 else 0 for sample in
    ↪    received_signal])

def hamming_decoding(encoded_bits, m=4):
    '''
    Perform Hamming (7,4) decoding.
    :param encoded_bits: Encoded bit stream.
    :param m: Number of data bits in one block.
    :return: Decoded data bits.
    '''
    n = 2**m - 1
    k = n - m
    # Parity-check matrix
    H = np.array([
        [1, 1, 1, 0, 1, 0, 0],
        [1, 0, 1, 1, 0, 1, 0],
        [0, 1, 1, 1, 0, 0, 1]
    ])
    syndrome_table = {
        (0, 0, 0): np.array([0, 0, 0, 0, 0, 0, 0]),
        (0, 0, 1): np.array([0, 0, 0, 0, 0, 0, 1]),
        (0, 1, 0): np.array([0, 0, 0, 0, 0, 1, 0]),
        (0, 1, 1): np.array([0, 0, 1, 0, 0, 0, 0]),
        (1, 0, 0): np.array([0, 0, 0, 1, 0, 0, 0]),
        (1, 0, 1): np.array([1, 0, 0, 0, 0, 0, 0]),
        (1, 1, 0): np.array([0, 1, 0, 0, 0, 0, 0]),
```

```
    (1, 1, 1): np.array([0, 0, 0, 0, 1, 0, 0]),
}
encoded_bits = np.array(encoded_bits).reshape(-1, n)
decoded_data = []
for block in encoded_bits:
    syndrome = np.mod(np.dot(H, block), 2)
    corrected_block = np.mod(block +
    ↪ syndrome_table[tuple(syndrome)], 2)
    decoded_data.append(corrected_block[:k])
return np.concatenate(decoded_data)

# Example usage
bit_stream = [0, 1, 1, 0, 1, 0, 0, 1]
modulated_signal = bpsk_modulation(bit_stream)
noisy_signal = add_awgn_noise(modulated_signal, snr=10)
demodulated_bits = bpsk_demodulation(noisy_signal)
encoded_bits = hamming_encoding(bit_stream)
decoded_bits = hamming_decoding(encoded_bits)

print("Original Bits:", bit_stream)
print("Modulated Signal:", modulated_signal)
print("Noisy Signal:", noisy_signal)
print("Demodulated Bits:", demodulated_bits.tolist())
print("Encoded Bits:", encoded_bits.tolist())
print("Decoded Bits:", decoded_bits.tolist())
```

This code implements several core functions necessary for simulating a basic satellite communication system:

- `bpsk_modulation` function converts a binary bit stream into a modulated BPSK signal.

- `add_awgn_noise` simulates the effect of channel noise using the AWGN model to degrade the signal based on a given SNR.

- `hamming_encoding` generates a Hamming (7,4) encoded bit stream from input data for error correction.

- `bpsk_demodulation` retrieves binary data from a received BPSK signal by applying a simple threshold rule.

- `hamming_decoding` corrects bit errors using the Hamming code, extracting the original data from the encoded stream.

These elements form the basis of communication methods such as encoding, modulation, and demodulation in a satellite communications system.

Chapter 19

Re-Entry Simulation and Ablation Modeling Algorithms

Below is a Python code snippet that details the computational elements for simulating spacecraft re-entry dynamics and thermal protection system ablation. This includes modeling of the re-entry trajectory, heat flux calculations, and ablation simulation using a simplified heat balance approach.

```python
import numpy as np

def re_entry_trajectory(v_initial, mass, angle, time_step,
↪ num_steps):
    '''
    Simulate the trajectory of a re-entering spacecraft.
    :param v_initial: Initial velocity (m/s).
    :param mass: Spacecraft mass (kg).
    :param angle: Re-entry angle (radians).
    :param time_step: Simulation time step (s).
    :param num_steps: Number of simulation steps.
    :return: Arrays of velocity and altitude over time.
    '''

    velocity = v_initial
    altitude = 120000  # Starting altitude in meters (e.g., 120 km)
    g = 9.81  # Gravity (m/s^2)
    velocities, altitudes = [velocity], [altitude]

    for _ in range(num_steps):
        drag = 0.5 * 1.225 * velocity**2 * 2.5 / mass
```

```python
        gravity_effect = g * np.sin(angle)
        velocity -= (drag + gravity_effect) * time_step
        altitude -= velocity * time_step

        velocities.append(velocity)
        altitudes.append(altitude)

        if altitude < 0:
            break

    return np.array(velocities), np.array(altitudes)

def heat_flux_calculation(velocity, v_ref=7800, heat_flux_ref=45e4):
    '''
    Calculate the heat flux experienced during re-entry.
    :param velocity: Current spacecraft velocity (m/s).
    :param v_ref: Reference velocity for heat flux calculation
    ↪  (m/s).
    :param heat_flux_ref: Reference heat flux (W/m^2).
    :return: Heat flux in W/m^2.
    '''

    return heat_flux_ref * (velocity / v_ref)**3

def ablation_simulation(heat_flux, duration,
↪    ablation_rate_coeff=1e-4):
    '''
    Simulate the ablation process for the thermal protection system.
    :param heat_flux: Heat flux (W/m^2).
    :param duration: Duration of heat exposure (s).
    :param ablation_rate_coeff: Ablation rate coefficient
    ↪  (kg/(s·m^2)).
    :return: Total ablated mass (kg).
    '''

    return ablation_rate_coeff * heat_flux * duration

# Simulation parameters
v_initial = 7800  # Initial velocity in m/s
mass = 2000  # Spacecraft mass in kg
angle = np.radians(5.5)  # Entry angle in radians
time_step = 1  # Time step in seconds
num_steps = 500  # Number of simulation steps

# Perform re-entry trajectory simulation
velocities, altitudes = re_entry_trajectory(v_initial, mass, angle,
↪    time_step, num_steps)

# Calculate heat flux and ablation over the trajectory
total_ablation = 0
for velocity in velocities:
    heat_flux = heat_flux_calculation(velocity)
    total_ablation += ablation_simulation(heat_flux, time_step)
```

```
print("Total Ablation (kg):", total_ablation)
```

This code includes several functions essential for spacecraft re-entry dynamics simulation:

- `re_entry_trajectory` simulates the re-entry trajectory, considering gravitational and drag forces influenced by re-entry angle.

- `heat_flux_calculation` calculates the heat flux based on current velocity, utilizing reference conditions.

- `ablation_simulation` computes the mass loss due to ablation based on heat flux and time duration.

The final section of the code performs the simulation over specified parameters, calculating the total ablation experienced during the re-entry phase.

Chapter 20

Aeroservoelastic Control Algorithms

Below is a Python code snippet for an aeroservoelastic analysis, integrating aerodynamic, structural, and control system dynamics to simulate the complex interactions in such systems.

```python
import numpy as np
from scipy.integrate import solve_ivp

def aerodynamic_forces(velocity, angle_of_attack):
    '''
    Calculate aerodynamic forces acting on the structure.
    :param velocity: Flight velocity.
    :param angle_of_attack: Angle of attack in radians.
    :return: Lift and drag forces.
    '''
    # Simplified linear model for demo purposes
    CL = 2 * np.pi * angle_of_attack
    CD = 0.02 + 0.1 * CL**2
    lift = 0.5 * velocity**2 * CL
    drag = 0.5 * velocity**2 * CD
    return lift, drag

def structural_dynamics(t, state, control_input):
    '''
    Equations of motion for the structural dynamics.
    :param t: Time variable.
    :param state: Current state [angle, angular velocity, structural
    ↪ deformation].
    :param control_input: Control force.
    :return: Derivatives of state.
    '''
```

```
        angle, angular_velocity, deformation = state
        damping_coefficient = 0.05
        stiffness = 10.0
        structural_force = -stiffness * deformation
        lift, drag = aerodynamic_forces(angular_velocity, angle)
        angle_dot = angular_velocity
        angular_velocity_dot = (lift + structural_force -
        ↪   damping_coefficient * angular_velocity + control_input) /
        ↪   100
        deformation_dot = angular_velocity  # Simple model for
        ↪   illustration
        return [angle_dot, angular_velocity_dot, deformation_dot]

def control_system(state):
    '''
    Control system to maintain desired aeroservoelastic performance.
    :param state: Current state [angle, angular velocity,
    ↪   deformation].
    :return: Control force.
    '''
    desired_angle = 0.0
    kp = 10.0
    control_force = kp * (desired_angle - state[0])
    return control_force

# Simulate the aeroservoelastic system
initial_conditions = [0.1, 0.0, 0.0]  # Initial angle, angular
↪   velocity, deformation
time_span = [0, 10]  # 10 seconds simulation

result = solve_ivp(
    lambda t, y: structural_dynamics(t, y, control_system(y)),
    time_span,
    initial_conditions,
    t_eval=np.linspace(0, 10, 100)  # Evaluate at 100 time points
)

# Plotting
import matplotlib.pyplot as plt

plt.figure(figsize=(12, 5))
plt.plot(result.t, result.y[0], label='Angle')
plt.plot(result.t, result.y[1], label='Angular Velocity')
plt.plot(result.t, result.y[2], label='Deformation')
plt.xlabel('Time (s)')
plt.ylabel('State Variables')
plt.title('Aeroservoelastic System Dynamics')
plt.legend()
plt.grid(True)
plt.show()
```

This code integrates core components necessary for aeroservoelastic analysis:

- `aerodynamic_forces` function computes lift and drag forces based on angle and velocity.

- `structural_dynamics` defines the structural dynamics' equations of motion with aerodynamic interactions.

- `control_system` implements a simple proportional controller to manage the aeroservoelastic responses.

- Simulation uses `scipy.integrate.solve_ivp` to solve the dynamic system over a specified time span, providing insights into the system's behavior.

The final block of code includes a simple demonstration of these interactions by plotting the system's angle, angular velocity, and deformation over time, illustrating the complex dynamics in aeroservoelastic systems.

Chapter 21

Computational Electromagnetics for Aerospace Applications

Below is a Python code snippet that encompasses the core computational elements for simulating electromagnetic fields in aerospace contexts, specifically focusing on antenna design and electromagnetic interference.

```python
import numpy as np
import matplotlib.pyplot as plt

def calculate_e_field(d, f, I, N=1000):
    '''
    Calculate the electric field strength of a dipole antenna.
    :param d: Separation between antenna elements (meters).
    :param f: Frequency of operation (Hz).
    :param I: Current in the antenna (Amperes).
    :param N: Number of points to calculate over the space.
    :return: Electric field magnitude over theta.
    '''
    c = 3e8  # speed of light (m/s)
    wavelength = c / f
    k = 2 * np.pi / wavelength  # wave number
    theta = np.linspace(0, 2 * np.pi, N)
    e_field = I * np.sin(k * d * np.cos(theta)) / (4 * np.pi * d *
    ↪ np.sin(theta))
    return theta, np.abs(e_field)

def plot_radiation_pattern(d, f, I):
    '''
```

```python
    Plot the radiation pattern of a dipole antenna.
    :param d: Separation between antenna elements (meters).
    :param f: Frequency of operation (Hz).
    :param I: Current in the antenna (Amperes).
    '''
    theta, e_field = calculate_e_field(d, f, I)
    plt.polar(theta, e_field)
    plt.title('Antenna Radiation Pattern')
    plt.show()

def calculate_emi(receiving_antenna_gain,
↪    transmitting_antenna_power, distance):
    '''
    Calculate electromagnetic interference (EMI) based on free-space
    ↪ path loss.
    :param receiving_antenna_gain: Gain of the receiving antenna
    ↪ (dimensionless).
    :param transmitting_antenna_power: Transmitted power (Watts).
    :param distance: Distance between antennas (meters).
    :return: EMI power at the receiving antenna (Watts).
    '''
    c = 3e8  # speed of light (m/s)
    wavelength = c / (f)  # using global frequency
    path_loss = (4 * np.pi * distance / wavelength) ** 2
    receiving_power = transmitting_antenna_power *
    ↪   receiving_antenna_gain / path_loss
    return receiving_power

# Example Parameters
d = 2.0  # Separation (m)
f = 1e9  # Frequency (Hz)
I = 1.0  # Current (A)

print("Simulating Antenna Radiation Pattern:")
plot_radiation_pattern(d, f, I)

receiving_antenna_gain = 10  # Gain (dimensionless)
transmitting_antenna_power = 50  # Power (W)
distance = 100  # Distance (m)

emi_power = calculate_emi(receiving_antenna_gain,
↪   transmitting_antenna_power, distance)
print(f"Calculated EMI Power at receiving antenna: {emi_power:.4f}
↪   W")
```

This code defines several key functions for evaluating electromagnetic simulation aspects within aerospace applications:

- `calculate_e_field` function computes the electric field strength of a dipole antenna based on its geometric properties, frequency, and current.

68

- `plot_radiation_pattern` visualizes the antenna's radiation pattern using polar plots to illustrate field strength variations around the dipole.

- `calculate_emi` function computes the electromagnetic interference power received by another antenna due to free-space path loss and other parameters.

The final block of code utilizes these functions to simulate and visualize the radiation pattern of a dipole antenna and calculate the electromagnetic interference power for given conditions.

Chapter 22

Adaptive Control Algorithms for Aerospace Vehicles

Below is a Python code snippet that focuses on adaptive control techniques for aerospace applications, illustrating how to design a controller that adapts to changing flight conditions and system uncertainties.

```python
import numpy as np

class AdaptiveController:
    def __init__(self, initial_gain, adaptation_rate):
        '''
        Initialize the adaptive controller.
        :param initial_gain: Initial controller gain.
        :param adaptation_rate: Rate at which the controller adapts
        ↪   to changes.
        '''
        self.gain = initial_gain
        self.adaptation_rate = adaptation_rate

    def update_gain(self, error, uncertainty):
        '''
        Update the controller gain based on the error and system
        ↪   uncertainty.
        :param error: The current error in system performance.
        :param uncertainty: The measure of uncertainty in the
        ↪   system.
        '''
        adjustment = self.adaptation_rate * error * uncertainty
```

```
        self.gain += adjustment

    def control_signal(self, measured_output, desired_output):
        '''
        Compute the control signal based on the measured and desired
        ↳  output.
        :param measured_output: The current output of the system.
        :param desired_output: The target output of the system.
        :return: Control signal to apply to the system.
        '''
        error = desired_output - measured_output
        control_signal = self.gain * error
        # For demonstration, we're assuming uncertainty measure as a
        ↳  scaled error
        uncertainty = np.abs(error) * 0.1
        self.update_gain(error, uncertainty)

        return control_signal

# Example usage of the AdaptiveController:
adaptive_ctrl = AdaptiveController(initial_gain=1.0,
↳  adaptation_rate=0.01)
desired_output = 10.0
measured_output = 7.5

control = adaptive_ctrl.control_signal(measured_output,
↳  desired_output)
print("Control Signal:", control)
print("Updated Gain:", adaptive_ctrl.gain)
```

This code defines a simple adaptive control system:

- **AdaptiveController** class initializes with an initial gain and
  an adaptation rate parameter, allowing the gains to be modified during operation.

- **update_gain** method adjusts the gain dynamically based on
  the error and an uncertainty factor, here exemplified as a
  function of the error itself.

- **control_signal** calculates the required control input to drive
  the system toward the desired state, while also adjusting the
  gain adaptively.

The example demonstrates how the **AdaptiveController** would
be used to compute control signals while continuously adjusting its
gain in response to changing system conditions and uncertainties.

Chapter 23

Machine Learning Algorithms in Aerospace Engineering

Below is a Python code snippet that encompasses the core computational elements of using machine learning algorithms for predictive maintenance, anomaly detection, and performance optimization, demonstrating data preparation, model training, and validation.

```python
import numpy as np
import pandas as pd
from sklearn.model_selection import train_test_split
from sklearn.ensemble import RandomForestClassifier
from sklearn.metrics import classification_report, accuracy_score
import matplotlib.pyplot as plt

# Sample data generation for demonstration
def generate_sample_data(num_samples=1000):
    '''
    Generates synthetic data representing sensor readings for
    ↪   workflows.
    :param num_samples: Number of data points.
    :return: DataFrame with features and labels.
    '''

    np.random.seed(42)
    temps = np.random.normal(loc=70, scale=10, size=num_samples)
    pressures = np.random.normal(loc=30, scale=5, size=num_samples)
    vibrations = np.random.normal(loc=5, scale=1, size=num_samples)
```

```python
    # Maintenance flag set to 1 when certain conditions are met
    ↪ (simulation)
    faults = (temps > 80) | (pressures < 25) | (vibrations > 7)
    labels = np.where(faults, 1, 0)

    data = pd.DataFrame({'temperature': temps, 'pressure':
    ↪ pressures, 'vibration': vibrations, 'maintenance_needed':
    ↪ labels})
    return data

# Data preprocessing
def preprocess_data(df):
    '''
    Splits data into training and test sets.
    :param df: Input DataFrame.
    :return: Split datasets.
    '''
    X = df[['temperature', 'pressure', 'vibration']]
    y = df['maintenance_needed']
    return train_test_split(X, y, test_size=0.25, random_state=42)

# Model training
def train_model(X_train, y_train):
    '''
    Trains a Random Forest classifier for predictive maintenance.
    :param X_train: Training features.
    :param y_train: Training labels.
    :return: Trained model.
    '''
    clf = RandomForestClassifier(n_estimators=100, random_state=42)
    clf.fit(X_train, y_train)
    return clf

# Main function to execute the machine learning pipeline
def execute_ml_pipeline():
    '''
    Executes the machine learning pipeline for predictive
    ↪ maintenance.
    :return: None.
    '''
    data = generate_sample_data()
    X_train, X_test, y_train, y_test = preprocess_data(data)
    model = train_model(X_train, y_train)

    # Model validation
    predictions = model.predict(X_test)
    print(classification_report(y_test, predictions))
    print(f"Accuracy: {accuracy_score(y_test, predictions):.2f}")

    # Feature importance visualization
    feature_importances = model.feature_importances_
    plt.bar(X_train.columns, feature_importances)
    plt.xlabel('Features')
```

```
plt.ylabel('Importance')
plt.title('Feature Importance for Predictive Maintenance Model')
plt.show()

# Execute the pipeline
execute_ml_pipeline()
```

This code defines several key components necessary for implementing machine learning algorithms for predictive maintenance:

- The `generate_sample_data` function simulates sensor data generating conditions that trigger maintenance needs.

- `preprocess_data` splits the dataset into training and testing subsets to prepare for model training and validation.

- `train_model` is responsible for training a Random Forest classifier to predict maintenance needs based on input features.

- `execute_ml_pipeline` ties the workflow, executing the data preparation, model training, evaluation, and visualization.

The pipeline demonstrates the use of synthetic data to validate a predictive model, presenting validation results and visualizing feature importance.

Chapter 24

Computational Aeroacoustics Algorithms

Below is a Python code snippet that demonstrates the core computational elements required to model and analyze noise generated by aerospace vehicles, particularly focusing on jet noise and rotorcraft acoustics.

```python
import numpy as np
import matplotlib.pyplot as plt

def calculate_jet_noise(frequency, velocity, density, diameter):
    '''
    Compute jet noise level based on input parameters.
    :param frequency: Frequency of interest (Hz).
    :param velocity: Jet velocity (m/s).
    :param density: Air density (kg/m^3).
    :param diameter: Nozzle diameter (m).
    :return: Noise level (dB).
    '''
    # Assuming a basic empirical model for demonstration
    reference_pressure = 20e-6  # Reference pressure in air (Pa)
    Strouhal_number = frequency * diameter / velocity
    SPL = 10 * np.log10((density * velocity**2) /
    ↪   (reference_pressure**2)) + 10 * np.log10(Strouhal_number)

    return SPL

def rotorcraft_acoustics(power, rotor_speed, blade_length,
↪   observer_distance):
```

```
    '''
    Compute rotorcraft acoustic signature.
    :param power: Power in kilowatts (kW).
    :param rotor_speed: Rotor speed in revolutions per minute (RPM).
    :param blade_length: Rotor blade length (m).
    :param observer_distance: Distance to observer (m).
    :return: Acoustic level (dB).
    '''
    # Simplistic model considering distance and power
    reference_distance = 1  # Reference distance (m)
    SPL = 20 * np.log10(power) + 10 * np.log10(rotor_speed) - 20 *
    ↪  np.log10(observer_distance / reference_distance)

    return SPL

def plot_noise_profile(frequencies, SPL_values):
    '''
    Plot the noise level profile over a range of frequencies.
    :param frequencies: Frequencies (Hz) over which to calculate
    ↪  SPL.
    :param SPL_values: Corresponding sound pressure levels (dB).
    '''
    plt.figure(figsize=(10, 5))
    plt.plot(frequencies, SPL_values, 'b-', label='Jet Noise')
    plt.xlabel('Frequency (Hz)')
    plt.ylabel('Sound Pressure Level (dB)')
    plt.title('Noise Profile')
    plt.legend()
    plt.grid(True)
    plt.show()

# Parameters for demonstration
frequencies = np.linspace(100, 10000, 100)
velocity = 340  # Jet velocity (m/s)
density = 1.225  # Air density (kg/m^3)
diameter = 0.5  # Nozzle diameter (m)

# Calculate jet noise for a range of frequencies
SPL_values = [calculate_jet_noise(f, velocity, density, diameter)
↪  for f in frequencies]

# Plot the jet noise profile
plot_noise_profile(frequencies, SPL_values)

# Compute rotorcraft acoustic signature as an example
power = 1000  # Power in kilowatts
rotor_speed = 500  # Rotor speed in RPM
blade_length = 10  # Rotor blade length in meters
observer_distance = 100  # Distance to observer in meters

rotorcraft_SPL = rotorcraft_acoustics(power, rotor_speed,
↪  blade_length, observer_distance)
```

76

This code snippet provides functions to model noise generation from jet engines and rotorcraft, visualize the noise profile, and calculate the acoustic signature:

- `calculate_jet_noise` computes the sound pressure level (SPL) for jet noise using the Strouhal number and empirical relationships.

- `rotorcraft_acoustics` evaluates the acoustic level of a rotorcraft based on factors like power, rotor speed, and observer distance.

- `plot_noise_profile` visualizes the noise profile across different frequencies, showcasing the SPL values.

The final section of the code demonstrates how to use these elements by calculating and plotting the noise profile for specified parameters and printing the rotorcraft acoustic level.

Chapter 25

High-Order Numerical Methods for CFD

Below is a Python code snippet that demonstrates high-order discretization schemes using spectral and discontinuous Galerkin methods typically utilized for enhancing CFD accuracy.

```python
import numpy as np

def spectral_method_1d(f, num_points, domain):
    '''
    Implementation of a simple spectral method for 1D function
    ↪ approximation.
    :param f: A function to be approximated.
    :param num_points: Number of points for discretization.
    :param domain: Tuple containing the start and end of the domain
    ↪ interval.
    :return: Approximated values using spectral method.
    '''

    # Create an array of points in the domain using Chebyshev nodes
    x = np.cos(np.pi * np.arange(num_points) / (num_points - 1))
    x = 0.5 * (domain[1] - domain[0]) * (x + 1) + domain[0]

    # Evaluate the function at the Chebyshev nodes
    f_values = f(x)

    # Compute the Chebyshev coefficients for the spectral expansion
    c = np.zeros(num_points)
    for k in range(num_points):
        c[k] = (2 / num_points) * sum(f_values * np.cos(k *
        ↪ np.arccos(-1 + 2 * np.arange(num_points) / (num_points -
        ↪ 1))))
        if k == 0 or k == num_points - 1:
```

```python
        c[k] *= 0.5

    # Create a finer grid for plotting the spectral approximation
    x_fine = np.linspace(domain[0], domain[1], 100)

    # Evaluate the spectral expansion at the finer grid points
    f_approx = np.zeros_like(x_fine)
    for k in range(num_points):
        f_approx += c[k] * np.cos(k * np.arccos((2 * x_fine -
        ↪   domain[0] - domain[1]) / (domain[1] - domain[0])))

    return x_fine, f_approx

def discontinuous_galerkin_1d(f, num_elements, basis_order, domain):
    '''
    Simplified Discontinuous Galerkin method example for 1D
    ↪   advection equation.
    :param f: Initial function.
    :param num_elements: Number of mesh elements.
    :param basis_order: Order of polynomial basis.
    :param domain: Tuple of domain interval.
    :return: Nodal values as DG solution after a timestep.
    '''
    h = (domain[1] - domain[0]) / num_elements
    nodes = np.linspace(domain[0], domain[1], num_elements *
    ↪   (basis_order + 1))

    # Initialize basis functions (Legendre polynomials)
    def legendre_basis(x, degree):
        '''Compute Legendre polynomial up to given degree'''
        if degree == 0:
            return np.ones_like(x)
        elif degree == 1:
            return x
        else:
            P0, P1 = np.ones_like(x), x
            for n in range(2, degree + 1):
                P = ((2 * n - 1) * x * P1 - (n - 1) * P0) / n
                P0, P1 = P1, P
            return P

    # Fill local basis function matrix
    local_basis = np.zeros((basis_order + 1, basis_order + 1))
    for i in range(basis_order + 1):
        nodes_b = np.linspace(-1, 1, basis_order + 1)
        local_basis[i, :] = legendre_basis(nodes_b, i)

    # Initial condition projection
    u = np.zeros_like(nodes)
    for i in range(num_elements):
        xi = np.linspace(domain[0] + i * h, domain[0] + (i + 1) * h,
        ↪   basis_order + 1)
        u[i * (basis_order + 1):(i + 1) * (basis_order + 1)] = f(xi)
```

79

```
    # Here we typically solve the DG equation (omitted for brevity)
    # This example assumes a simplistic constant velocity advection

    return nodes, u

# Example usage for both methods
def example_function(x):
    return np.sin(np.pi * x)

# Spectral method
domain = (0, 1)
num_points = 20
x_fine, f_approx = spectral_method_1d(example_function, num_points,
↪   domain)

# Discontinuous Galerkin method
num_elements = 10
basis_order = 3
nodes, u = discontinuous_galerkin_1d(example_function, num_elements,
↪   basis_order, domain)

# Visualization would occur here (e.g., using Matplotlib) to
↪   showcase the approximations.
# The above code is groundwork for such implementation.
```

This code defines essential functions for implementing high-order discretization schemes in CFD:

- The `spectral_method_1d` function approximates a 1D function using a spectral method, employing Chebyshev polynomials for high accuracy.

- The `discontinuous_galerkin_1d` function demonstrates a basic implementation of the Discontinuous Galerkin method on a 1D domain using Legendre polynomial basis functions.

Both methods represent advanced techniques designed to enhance accuracy in computational fluid dynamics simulations, particularly beneficial for capturing complex flow dynamics accurately.

Chapter 26

Autonomous Flight Control Algorithms

Below is a Python code snippet that encompasses the core computational elements enabling autonomous operation of aerospace vehicles, focusing particularly on sense-and-avoid and path planning functionalities.

```python
import numpy as np
import matplotlib.pyplot as plt
from scipy.spatial import KDTree

def sense_and_avoid(vehicle_position, obstacles, safe_distance):
    '''
    Determine if there are nearby obstacles requiring avoidance and
    ↪    return avoidance vector.
    :param vehicle_position: Current position of the vehicle (x, y).
    :param obstacles: List of obstacle positions [(x1, y1), (x2,
    ↪    y2), ...].
    :param safe_distance: Safe distance threshold to avoid
    ↪    collision.
    :return: Avoidance vector (dx, dy).
    '''
    avoidance_vector = np.array([0.0, 0.0])
    for obs in obstacles:
        distance = np.linalg.norm(np.array(obs) -
        ↪    np.array(vehicle_position))
        if distance < safe_distance:
            diff = np.array(vehicle_position) - np.array(obs)
            avoidance_vector += diff / distance * (safe_distance -
            ↪    distance)
    return avoidance_vector
```

```python
def path_planning(start, goal, obstacles, step_size=1.0):
    '''
    Plan a path from start to goal avoiding obstacles using a simple
    ↪ potential field.
    :param start: Starting position of the vehicle (x, y).
    :param goal: Goal position of the vehicle (x, y).
    :param obstacles: List of obstacle positions [(x1, y1), (x2,
    ↪ y2), ...].
    :param step_size: Movement step size.
    :return: List of path points [(x1, y1), (x2, y2), ...].
    '''
    path = [start]
    current_position = np.array(start)
    goal_position = np.array(goal)

    while np.linalg.norm(current_position - goal_position) >
    ↪ step_size:
        towards_goal = (goal_position - current_position)
        towards_goal = towards_goal / np.linalg.norm(towards_goal)
        avoidance = sense_and_avoid(current_position, obstacles,
        ↪ safe_distance=2.0)

        move = towards_goal + avoidance
        move = move / np.linalg.norm(move) * step_size
        current_position = current_position + move

        path.append(tuple(current_position))

    return path

# Visualizing the planned path
def plot_path(start, goal, obstacles, path):
    '''
    Create a plot showing the path from start to goal and obstacles.
    :param start: Starting position of the vehicle.
    :param goal: Goal position of the vehicle.
    :param obstacles: Obstacle positions.
    :param path: Planned path points.
    '''
    plt.figure()
    plt.plot(*zip(*[start, goal]), 'go', label='Start/Goal')
    plt.plot(*zip(*obstacles), 'ro', label='Obstacles')
    if path:
        plt.plot(*zip(*path), 'b-', label='Path')
    plt.legend()
    plt.xlabel('X')
    plt.ylabel('Y')
    plt.title('Autonomous Path Planning')
    plt.grid(True)
    plt.show()

# Example usage
start_pos = (0, 0)
```

```
goal_pos = (10, 10)
obstacle_list = [(4, 5), (5, 6), (6, 5)]

planned_path = path_planning(start_pos, goal_pos, obstacle_list)
plot_path(start_pos, goal_pos, obstacle_list, planned_path)
```

This code defines key functionalities necessary for the autonomous operation of aerospace vehicles:

- `sense_and_avoid` function checks for nearby obstacles and calculates an avoidance vector to steer the vehicle away from potential collisions.

- `path_planning` generates a path from a start to a goal position while avoiding obstacles, accomplished through a potential field-based approach.

- `plot_path` visually represents the planned path, start, goal, and obstacles in a 2D plot, aiding in the analysis and validation of autonomous path planning.

The example demonstrates a simple scenario of path planning with specified start and goal positions amidst a set of obstacles.

Chapter 27

Spacecraft Formation Flying Algorithms

Below is a Python code snippet that encompasses the core computational elements for controlling multiple spacecraft flying in formation, focusing on relative navigation and control. The implementation includes a simulation of spacecraft dynamics, a control law for maintaining formation, and visualization tools for analysis.

```python
import numpy as np
import matplotlib.pyplot as plt

def spacecraft_dynamics(position, velocity, control_input, dt):
    '''
    Simulate the dynamics of a spacecraft.
    :param position: Current position of the spacecraft.
    :param velocity: Current velocity of the spacecraft.
    :param control_input: Control force applied to the spacecraft.
    :param dt: Time step for the simulation.
    :return: Updated position and velocity.
    '''
    acceleration = control_input
    new_velocity = velocity + acceleration * dt
    new_position = position + new_velocity * dt
    return new_position, new_velocity

def control_law(target_position, current_position,
    current_velocity):
    '''
    Compute the control input to maintain formation.
    :param target_position: Desired position of the spacecraft.
    :param current_position: Current position of the spacecraft.
```

```python
    :param current_velocity: Current velocity of the spacecraft.
    :return: Control input for the spacecraft.
    '''
    k_p = 0.1  # Proportional gain
    k_d = 0.05  # Derivative gain
    position_error = target_position - current_position
    velocity_error = -current_velocity
    control_input = k_p * position_error + k_d * velocity_error
    return control_input


def simulate_formation_flight(num_spacecraft, timesteps, dt):
    '''
    Simulate the formation flight of multiple spacecraft.
    :param num_spacecraft: Number of spacecraft in the formation.
    :param timesteps: Number of time steps to simulate.
    :param dt: Size of each time step.
    :return: Trajectories of all spacecraft.
    '''
    # Initialize positions and velocities
    positions = np.random.rand(num_spacecraft, 2) * 100
    velocities = np.zeros((num_spacecraft, 2))
    trajectories = [positions.copy()]

    # Define target formation
    target_positions = np.linspace(start=[0, 0], stop=[100, 100],
    ↪   num=num_spacecraft)

    # Simulate over time
    for _ in range(timesteps):
        for i in range(num_spacecraft):
            control_input = control_law(target_positions[i],
            ↪   positions[i], velocities[i])
            positions[i], velocities[i] =
            ↪   spacecraft_dynamics(positions[i], velocities[i],
            ↪   control_input, dt)
        trajectories.append(positions.copy())

    return np.array(trajectories)


def plot_trajectories(trajectories):
    '''
    Plot the trajectories of the spacecraft during the simulation.
    :param trajectories: Array of spacecraft positions over time.
    '''
    num_spacecraft = trajectories.shape[1]
    for i in range(num_spacecraft):
        plt.plot(trajectories[:, i, 0], trajectories[:, i, 1],
        ↪   label=f'Spacecraft {i+1}')
    plt.xlabel('X Position')
    plt.ylabel('Y Position')
    plt.title('Spacecraft Formation Flight Trajectories')
    plt.legend()
    plt.show()
```

```
# Simulation parameters
num_spacecraft = 5
timesteps = 100
dt = 0.1

# Perform simulation
trajectories = simulate_formation_flight(num_spacecraft, timesteps,
↪  dt)

# Plot results
plot_trajectories(trajectories)
```

This code defines several key functions necessary for simulating and controlling multiple spacecraft flying in formation:

- `spacecraft_dynamics` function simulates the dynamics of a spacecraft given its current state and control inputs.

- `control_law` computes the control input required to maintain the desired formation, employing a simple proportional-derivative controller.

- `simulate_formation_flight` integrates the spacecraft dynamics over time, simulating the behavior of multiple spacecraft.

- `plot_trajectories` generates a visual representation of the simulated trajectories to aid in analysis.

The final block of code sets up the simulation parameters, executes the formation flight simulation, and visualizes the trajectory outcomes for inspection.

Chapter 28

Finite Volume Methods for Fluid Dynamics

Below is a Python code snippet that encompasses a finite volume discretization algorithm for solving fluid flow problems in complex aerospace geometries.

```python
import numpy as np

def initialize_geometry(domain_size, dx, dy):
    '''
    Initialize the computational grid for finite volume method.
    :param domain_size: Tuple of the domain size (width, height).
    :param dx: Grid spacing in the x-direction.
    :param dy: Grid spacing in the y-direction.
    :return: Mesh grid points in the x and y directions.
    '''
    nx, ny = domain_size
    x = np.linspace(0, nx * dx, nx)
    y = np.linspace(0, ny * dy, ny)
    return np.meshgrid(x, y)

def initialize_conditions(u_initial, v_initial, nx, ny):
    '''
    Initialize the velocity fields.
    :param u_initial: Initial condition for the u-velocity
    ↪   component.
    :param v_initial: Initial condition for the v-velocity
    ↪   component.
    :param nx: Number of grid points in the x-direction.
    :param ny: Number of grid points in the y-direction.
    :return: Initialized velocity fields u and v.
    '''
```

```python
    u = np.ones((nx, ny)) * u_initial
    v = np.ones((nx, ny)) * v_initial
    return u, v

def compute_fluxes(u, v, dx, dy, mu):
    '''
    Compute convective and diffusive fluxes.
    :param u: Velocity field in x-direction.
    :param v: Velocity field in y-direction.
    :param dx: Grid spacing in the x-direction.
    :param dy: Grid spacing in the y-direction.
    :param mu: Dynamic viscosity.
    :return: Convective and diffusive fluxes for momentum equations.
    '''
    # Example calculation of convective fluxes
    convective_u = u * (np.roll(u, -1, axis=1) - u) / dx
    convective_v = v * (np.roll(v, -1, axis=0) - v) / dy

    # Example calculation of diffusive fluxes
    diffusive_u = mu * (np.roll(u, -1, axis=1) - 2 * u + np.roll(u,
↪    1, axis=1)) / dx**2 + \
                mu * (np.roll(u, -1, axis=0) - 2 * u + np.roll(u,
↪        1, axis=0)) / dy**2
    diffusive_v = mu * (np.roll(v, -1, axis=0) - 2 * v + np.roll(v,
↪    1, axis=0)) / dy**2 + \
                mu * (np.roll(v, -1, axis=1) - 2 * v + np.roll(v,
↪        1, axis=1)) / dx**2

    return convective_u, convective_v, diffusive_u, diffusive_v

def update_velocity(u, v, dt, dx, dy, convective_u, convective_v,
↪  diffusive_u, diffusive_v, rho):
    '''
    Update velocity fields using finite volume method.
    :param u: Current velocity field in x-direction.
    :param v: Current velocity field in y-direction.
    :param dt: Time step size.
    :param dx: Grid spacing in the x-direction.
    :param dy: Grid spacing in the y-direction.
    :param convective_u: Convective fluxes for u.
    :param convective_v: Convective fluxes for v.
    :param diffusive_u: Diffusive fluxes for u.
    :param diffusive_v: Diffusive fluxes for v.
    :param rho: Fluid density.
    :return: Updated velocity fields u and v.
    '''
    u_new = u - dt * (convective_u + diffusive_u) / rho
    v_new = v - dt * (convective_v + diffusive_v) / rho
    return u_new, v_new

def main_simulation_loop(domain_size, dx, dy, dt, steps, u_initial,
↪  v_initial, mu, rho):
    '''
```

```
Run the main simulation loop for the fluid flow problem.
:param domain_size: Tuple of the domain size (width, height).
:param dx: Grid spacing in the x-direction.
:param dy: Grid spacing in the y-direction.
:param dt: Time step size.
:param steps: Number of simulation steps.
:param u_initial: Initial condition for the u-velocity
↪ component.
:param v_initial: Initial condition for the v-velocity
↪ component.
:param mu: Dynamic viscosity.
:param rho: Density of the fluid.
'''
x, y = initialize_geometry(domain_size, dx, dy)
u, v = initialize_conditions(u_initial, v_initial, *domain_size)

for _ in range(steps):
    conv_u, conv_v, diff_u, diff_v = compute_fluxes(u, v, dx,
    ↪ dy, mu)
    u, v = update_velocity(u, v, dt, dx, dy, conv_u, conv_v,
    ↪ diff_u, diff_v, rho)

return u, v

# Simulation parameters
domain_size = (100, 100)
dx = 0.01
dy = 0.01
dt = 0.001
steps = 100
u_initial = 1.0
v_initial = 0.0
mu = 0.01
rho = 1.0

# Run simulation
u_final, v_final = main_simulation_loop(domain_size, dx, dy, dt,
↪ steps, u_initial, v_initial, mu, rho)

# Print information
print("Final velocity fields computed.")
```

This code outlines several critical components for solving fluid flow problems using the finite volume method in a complex geometry typical of aerospace applications:

- initialize_geometry sets up the computational grid, creating a mesh based on the specified domain size and grid spacing.

- initialize_conditions initializes the velocity fields necessary for start of the simulation.

- `compute_fluxes` calculates both convective and diffusive fluxes essential for momentum equations using finite volume discretization.

- `update_velocity` updates the velocity fields over time using calculated fluxes and fluid properties.

- `main_simulation_loop` orchestrates the entire simulation process, utilizing previously defined functions to evolve the fluid flow system step-by-step.

The final output gives the velocity fields after the specified simulation steps, demonstrating the flow solution for the problem setup.

Chapter 29

Optimization Algorithms for Structural Design

Below is a Python code snippet that encompasses the core computational elements of optimizing structural components under weight, strength, and compliance constraints using a simple optimization framework with constraints on structural properties.

```python
import numpy as np
from scipy.optimize import minimize

def structural_objective(x):
    '''
    Objective function for structural optimization.
    In a real-world case, this could represent material usage, cost,
    ↪    etc.
    :param x: Design variables array.
    :return: Objective value.
    '''

    return np.sum(x**2)  # Placeholder for a real objective function

def weight_constraint(x, max_weight):
    '''
    Weight constraint function.
    :param x: Design variables array.
    :param max_weight: Maximum allowable weight.
    :return: Constraint value (should be <= 0).
    '''

    weight = np.sum(x)
    return weight - max_weight
```

```python
def compliance_constraint(x, max_compliance):
    '''
    Compliance (flexibility) constraint function.
    :param x: Design variables array.
    :param max_compliance: Maximum allowable compliance.
    :return: Constraint value (should be <= 0).
    '''
    compliance = np.prod(x)
    return compliance - max_compliance

def strength_constraint(x, required_strength):
    '''
    Strength constraint function.
    :param x: Design variables array.
    :param required_strength: Minimum required strength.
    :return: Constraint value (should be <= 0).
    '''
    strength = np.min(x)
    return required_strength - strength

# Design variables (initial guess)
x0 = np.array([1.0, 1.0, 1.0])

# Constraints setup
constraints = [
    {'type': 'ineq', 'fun': lambda x: -weight_constraint(x,
    ↪ max_weight=5.0)},
    {'type': 'ineq', 'fun': lambda x: -compliance_constraint(x,
    ↪ max_compliance=2.0)},
    {'type': 'ineq', 'fun': lambda x: -strength_constraint(x,
    ↪ required_strength=0.5)}
]

# Solve the optimization problem
result = minimize(structural_objective, x0, method='SLSQP',
↪ constraints=constraints)

print("Optimal design variables:", result.x)
print("Objective function value:", result.fun)
```

This code defines several key elements necessary for structural optimization under specified constraints:

- `structural_objective` function represents a simplified objective that could be expanded based on real-world criteria such as material cost.

- `weight_constraint` checks if the total weight is within an allowable limit.

- `compliance_constraint` ensures the design meets maximum compliance requirements.

- `strength_constraint` verifies that the structure's strength exceeds a required minimum.

- The optimization is performed using the `scipy.optimize.minimize` method with Sequential Least Squares Programming (SLSQP).

The final block of code demonstrates solving this optimization problem with a given initial guess and constraints.

Chapter 30

Computational Methods for Space Debris Modeling

Below is a Python code snippet that demonstrates a basic algorithm to track and predict the motion of space debris using numerical integration and predictions for collision avoidance.

```python
import numpy as np
from scipy.integrate import solve_ivp

def debris_motion_ODE(t, y):
    '''
    Defines the ordinary differential equations for the space debris
    ↪   motion model.
    :param t: Time variable.
    :param y: State vector [x, y, z, vx, vy, vz].
    :return: Derivative of the state vector.
    '''
    x, y, z, vx, vy, vz = y
    # Constants
    mu = 398600.0   # Earth's gravitational parameter, km^3/s^2

    # Equations of motion
    r3 = (x**2 + y**2 + z**2)**1.5
    ax = -mu * x / r3
    ay = -mu * y / r3
    az = -mu * z / r3
    return [vx, vy, vz, ax, ay, az]

def predict_debris_trajectory(initial_state, t_span):
```

```
'''
Predicts the trajectory of space debris using numerical
↪  integration.
:param initial_state: Initial state vector [x, y, z, vx, vy,
↪  vz].
:param t_span: Time span for the simulation [start_time,
↪  end_time].
:return: Time points, Trajectory solution array.
'''
sol = solve_ivp(debris_motion_ODE, t_span, initial_state,
↪  method='RK45', rtol=1e-6, atol=1e-9)
return sol.t, sol.y

def assess_collision_risk(debris_trajectory, threshold_distance):
    '''
    Assesses the risk of collision by checking the minimum distance
    ↪  to a threshold.
    :param debris_trajectory: Trajectory solution array.
    :param threshold_distance: Distance threshold for collision
    ↪  risk, km.
    :return: Boolean indicating collision risk.
    '''
    positions = debris_trajectory[:3, :]
    distances = np.linalg.norm(positions, axis=0)
    return np.any(distances < threshold_distance)

# Example usage
initial_state = [7000.0, 0.0, 0.0, 0.0, 7.5, 0.0]  # Initial
↪  position (km) and velocity (km/s)
t_span = [0.0, 86400.0]  # Simulate for one day (seconds)

time_points, trajectory = predict_debris_trajectory(initial_state,
↪  t_span)
collision_risk = assess_collision_risk(trajectory,
↪  threshold_distance=10.0)  # Threshold at 10 km

print("Time Points:", time_points)
print("Trajectory:", trajectory)
print("Collision Risk:", collision_risk)
```

This code provides a basic framework for space debris tracking and collision risk assessment:

- **debris_motion_ODE** defines the equations of motion based on gravitational dynamics.

- **predict_debris_trajectory** integrates these equations over time to predict the debris path.

- **assess_collision_risk** evaluates whether the predicted trajectory brings the debris within a critical distance threshold, indicating a collision risk.

The demo example sets an initial state vector for the debris and simulates its motion over one day, checking for collision risk.

Chapter 31

Fluid-Structure Interaction Algorithms

Below is a Python code snippet that demonstrates the interaction between fluid flows and structural responses using the Fluid-Structure Interaction (FSI) algorithm. This example includes a simulation of a simplified scenario where the fluid exerts pressure on a deformable structure. The code integrates methods for solving fluid dynamics using computational fluid dynamics (CFD) principles and structural dynamics using the finite element method (FEM).

```python
import numpy as np
from scipy.integrate import solve_ivp


def fluid_dynamics_solver(pressure_func, t_span, y0, c=0.5):
    '''
    Solves the fluid dynamics pressure effect over time on the
    ↪  structure.
    :param pressure_func: Function defining pressure over time.
    :param t_span: Time span for integration.
    :param y0: Initial state of the system.
    :param c: Fluid damping coefficient.
    :return: Solution containing time points and pressure effects.
    '''
    def fluid_ode(t, y):
        pressure = pressure_func(t)
        return c * pressure - y  # Sample dynamic model

    sol = solve_ivp(fluid_ode, t_span, y0, method='RK45', rtol=1e-5)
    return sol
```

```python
def structural_solver(displacement, pressure_effect, k=1.0, m=1.0):
    '''
    Computes the structural response to the fluid pressure using
    ↪   basic FEM.
    :param displacement: Initial displacement of structure.
    :param pressure_effect: Pressure effect from fluid dynamics.
    :param k: Stiffness of the structure.
    :param m: Mass of the structure.
    :return: Computed displacement considering pressure effect.
    '''
    return (pressure_effect / (k * m)) - displacement  # Simplified
    ↪   response


def pressure_function(t):
    '''
    Defines a pressure function varying over time.
    :param t: Time variable.
    :return: Calculated pressure value.
    '''
    return 10 * np.sin(t)  # Example sinusoidal pressure change


# Initialize simulation parameters
t_span = (0, 10)  # Time span for simulation
y0 = [0]  # Initial condition for fluid ODE
displacement_initial = 0.1  # Initial displacement of the structure

# Solve for fluid dynamics response
fluid_solution = fluid_dynamics_solver(pressure_function, t_span,
↪   y0)

# Solve for structural response over the fluid dynamics solution
displacement_response = [structural_solver(displacement_initial, p)
↪   for p in fluid_solution.y[0]]

# Print out results for each time step
for t, disp in zip(fluid_solution.t, displacement_response):
    print(f"Time: {t:.2f}, Displacement: {disp:.4f}")
```

This code defines several key functions demonstrating the Fluid-Structure Interaction (FSI) process through:

- `fluid_dynamics_solver` function uses a simple model to solve for the pressure exerted on a structure over time using the `pressure_func` provided.

- `structural_solver` is implemented to calculate the defor-

98

mation of the material based on pressure effects and predefined structural properties.

- `pressure_function` provides a basic example of how dynamic pressure might be modeled externally over time.

The overall integration provides a holistic view of fluid forces
interacting with a structural system, which is central to many
aerospace engineering challenges.

Chapter 32

Particle Methods in Aerospace Simulations

Below is a Python code snippet that represents the core computational elements for Smoothed Particle Hydrodynamics (SPH) simulation, useful for scenarios where traditional grid methods may not suffice.

```python
import numpy as np

class Particle:
    def __init__(self, position, velocity, mass, pressure, density):
        self.position = np.array(position)
        self.velocity = np.array(velocity)
        self.mass = mass
        self.pressure = pressure
        self.density = density
        self.force = np.zeros_like(self.position)

def sph_kernel(r, h):
    '''
    Compute the value of the SPH kernel.
    :param r: Distance between particles.
    :param h: Smoothing length.
    :return: Kernel value.
    '''
    q = r / h
    sigma = 1 / (np.pi * h**2)
    if q <= 1.0:
        return sigma * (1 - 3/2 * q**2 + 3/4 * q**3)
    elif 1.0 < q <= 2.0:
        return sigma * (1/4 * (2 - q)**3)
    else:
```

```
        return 0

def compute_density(particles, h):
    '''
    Compute the density for each particle.
    :param particles: List of particles.
    :param h: Smoothing length.
    '''

    for i, pi in enumerate(particles):
        density_sum = 0.0
        for j, pj in enumerate(particles):
            r = np.linalg.norm(pi.position - pj.position)
            density_sum += pj.mass * sph_kernel(r, h)
        pi.density = density_sum

def compute_pressure(particles, k, rest_density):
    '''
    Update pressure for each particle.
    :param particles: List of particles.
    :param k: Pressure constant.
    :param rest_density: Rest density of the fluid.
    '''

    for particle in particles:
        particle.pressure = k * (particle.density - rest_density)

def compute_forces(particles, h, viscosity):
    '''
    Compute forces on each particle considering pressure and
    ↪ viscosity.
    :param particles: List of particles.
    :param h: Smoothing length.
    :param viscosity: Viscosity coefficient.
    '''

    for i, pi in enumerate(particles):
        pi.force.fill(0)  # Reset force
        for j, pj in enumerate(particles):
            if i != j:
                r = np.linalg.norm(pi.position - pj.position)
                e_ij = (pi.position - pj.position) / (r + 1e-12)  #
                ↪ Unit vector with regularization
                pressure_term = -pj.mass * (pi.pressure +
                ↪ pj.pressure) / (2 * pj.density) * sph_kernel(r,
                ↪ h)
                viscosity_term = viscosity * pj.mass * (pj.velocity
                ↪ - pi.velocity) / pj.density * sph_kernel(r, h)
                pi.force += pressure_term * e_ij + viscosity_term

def update_particles(particles, dt):
    '''
    Update particle positions and velocities.
    :param particles: List of particles.
    :param dt: Time step.
    '''
```

```
    for particle in particles:
        particle.velocity += dt * particle.force / particle.density
        particle.position += dt * particle.velocity

# Example simulation setup
particles = [Particle(position=[0, 0], velocity=[0, 0], mass=1.0,
↪ pressure=0, density=0) for _ in range(100)]
h = 1.0
k = 1.0
rest_density = 1.0
viscosity = 0.1
dt = 0.01

# Simulation loop
for _ in range(1000):  # Run for 1000 timesteps
    compute_density(particles, h)
    compute_pressure(particles, k, rest_density)
    compute_forces(particles, h, viscosity)
    update_particles(particles, dt)
```

This code defines several key functions necessary for implementing the Smoothed Particle Hydrodynamics (SPH) method:

- `sph_kernel` computes the SPH kernel function, crucial for particle interaction weighting based on distance.

- `compute_density` evaluates the fluid density at each particle by summing contributions from neighboring particles.

- `compute_pressure` calculates the pressure at each particle using an equation of state.

- `compute_forces` determines the total force actuation on each particle by considering both pressure forces and viscous damping.

- `update_particles` integrates particle velocities and positions over time using the computed forces.

This simulation provides a basic framework for simulating fluid-like particle dynamics using SPH, suitable for scenarios where grid-based methods may become cumbersome or inefficient.

Chapter 33

Robust Control Algorithms for Aerospace Systems

Below is a Python code snippet that implements robust control algorithms to maintain performance in aerospace systems under the presence of uncertainties and disturbances, featuring a basic linear quadratic Gaussian (LQG) controller.

```python
import numpy as np
import scipy.linalg

def lqr(A, B, Q, R):
    '''
    Solve the continuous time LQR controller for a system described
    ↪  by matrices A, B
    :param A: State matrix
    :param B: Input matrix
    :param Q: State cost matrix
    :param R: Input cost matrix
    :return: Optimal gain matrix K
    '''
    # Solve the continuous time algebraic Riccati equation
    X = np.matrix(scipy.linalg.solve_continuous_are(A, B, Q, R))

    # Compute the LQR gain
    K = np.matrix(scipy.linalg.inv(R) @ (B.T @ X))

    return K

def kalman_filter(A, C, Q, R, measurements):
```

```python
    '''
    Implement a Kalman filter for state estimation
    :param A: State transition matrix
    :param C: Measurement matrix
    :param Q: Process noise covariance
    :param R: Measurement noise covariance
    :param measurements: Array of measurements
    :return: Estimated states
    '''
    n = A.shape[0]
    m = measurements.shape[1]

    # Initial guesses
    x_est = np.zeros((n, 1))
    P_est = np.eye(n)

    # Store the estimates
    estimated_states = []

    for z in measurements:
        # Prediction step
        x_pred = A @ x_est
        P_pred = A @ P_est @ A.T + Q

        # Update step
        y = z.reshape(m, 1) - (C @ x_pred)
        S = C @ P_pred @ C.T + R
        K = P_pred @ C.T @ np.linalg.inv(S)
        x_est = x_pred + (K @ y)
        P_est = (np.eye(n) - (K @ C)) @ P_pred

        # Save the estimated state
        estimated_states.append(x_est.flatten())

    return np.array(estimated_states)

def apply_control(K, x):
    '''
    Apply the control input based on LQR feedback
    :param K: LQR gain matrix
    :param x: Current state
    :return: Control input
    '''
    return -K @ x

# Example parameters for simulation
A = np.array([[1.1, 2.0], [0.3, 0.95]])
B = np.array([[0.1], [0.05]])
C = np.array([[1, 0]])
Q = np.eye(2)
R = np.array([[0.1]])
Q_kalman = np.eye(2) * 0.01
R_kalman = np.array([[0.1]])
```

104

```
K = lqr(A, B, Q, R)
measurements = np.array([[5 + np.random.randn(), 5 +
↪   np.random.randn()] for _ in range(10)])
estimated_states = kalman_filter(A, C, Q_kalman, R_kalman,
↪   measurements)

# Apply control based on the state estimates
for state in estimated_states:
    control_input = apply_control(K, np.array([state]).T)
    print("Control Input:", control_input)
```

This code defines several core functions for maintaining control under uncertainties:

- `lqr` function computes the optimal linear quadratic regulator gain for state feedback in continuous time systems.

- `kalman_filter` implements a Kalman filter for estimating the system state based on noisy measurements.

- `apply_control` determines the control input using the gain computed by the LQR function against the current state.

The code features an example setup with the calculation of the control strategy over a series of estimated states, ensuring the robustness of control despite the presence of noise and disturbances in the measurements.

Chapter 34

Computational Methods for Combustion Simulation

Below is a Python code snippet that encompasses the core computational elements for simulating combustion processes in engines, including chemical kinetics and flame modeling.

```python
import numpy as np
from scipy.integrate import odeint
import matplotlib.pyplot as plt

def reaction_rate(C, T, k0, Ea, R=8.314):
    '''
    Calculate the reaction rate using the Arrhenius equation.
    :param C: Concentration of the reactant.
    :param T: Temperature in Kelvin.
    :param k0: Pre-exponential factor.
    :param Ea: Activation energy in J/mol.
    :param R: Universal gas constant.
    :return: Reaction rate.
    '''
    k = k0 * np.exp(-Ea / (R * T))
    return k * C

def combustion_ode(C, t, T, k0, Ea):
    '''
    Define the system of ODEs for combustion kinetics.
    :param C: Concentration of the reactant.
    :param t: Time variable.
    :param T: Temperature in Kelvin.
```

```
    :param k0: Pre-exponential factor.
    :param Ea: Activation energy in J/mol.
    :return: Derivative of concentration with respect to time.
    '''
    dCdt = -reaction_rate(C, T, k0, Ea)
    return dCdt

# Initial conditions
C0 = 1.0   # Initial concentration of reactant
T = 1500   # Temperature in Kelvin
k0 = 1e10   # Pre-exponential factor
Ea = 80000   # Activation energy in J/mol
t = np.linspace(0, 0.01, 100)   # Time grid

# Solve ODE
C = odeint(combustion_ode, C0, t, args=(T, k0, Ea))

def plot_results(t, C):
    '''
    Plot the concentration over time.
    :param t: Time array.
    :param C: Concentration array.
    '''
    plt.figure()
    plt.plot(t, C, 'b', label='Concentration of Reactant')
    plt.xlabel('Time [s]')
    plt.ylabel('Concentration [mol/L]')
    plt.title('Combustion Kinetics')
    plt.legend(loc='best')
    plt.grid(True)
    plt.show()

# Plot results
plot_results(t, C)
```

This code defines several key functions necessary for simulating combustion processes:

- `reaction_rate` function computes the rate of a chemical reaction using the Arrhenius equation, which is fundamental to modeling reaction kinetics.

- `combustion_ode` sets up the ordinary differential equation (ODE) system representing the rate of change of concentration over time due to the combustion process.

- `plot_results` function generates plots to visualize the concentration of reactants over time, aiding in the analysis of combustion dynamics.

The final block of code runs a simulation and plots the concentration decrease over time for a simple combustion reaction, providing insights into the reaction kinetics and flame dynamics.

Chapter 35

Adaptive Mesh Refinement Algorithms

Below is a Python code snippet that demonstrates the implementation of adaptive mesh refinement (AMR) algorithms, which dynamically adapt computational meshes to capture solution features efficiently.

```python
import numpy as np
from matplotlib import pyplot as plt
from scipy.interpolate import griddata

# Define a function that represents a complex solution, e.g., a
↪ Gaussian
def solution_function(x, y):
    '''
    Complex solution function to be captured by the mesh.
    :param x: X positions.
    :param y: Y positions.
    :return: Solution value at given positions.
    '''
    return np.exp(-((x - 0.5) ** 2 + (y - 0.5) ** 2) / 0.01)

# Generate an initial coarse mesh
initial_points = np.random.rand(100, 2)
initial_values = solution_function(initial_points[:, 0],
↪ initial_points[:, 1])

# Adaptive mesh refinement function
def adaptive_mesh_refinement(points, values, refinement_threshold):
    '''
    Perform adaptive mesh refinement by adding points in regions
    ↪ with high error.
```

```
:param points: Array of points in the mesh.
:param values: Array of solution values at each point.
:param refinement_threshold: Error threshold for refinement.
:return: Refined mesh points and values.
'''
# Create a grid for interpolation
grid_x, grid_y = np.mgrid[0:1:100j, 0:1:100j]
grid_values = griddata(points, values, (grid_x, grid_y),
  ↪  method='cubic', fill_value=0.0)

# Calculate error estimate (simple gradient-based method here)
gradient = np.gradient(grid_values)
error_estimate = np.hypot(gradient[0], gradient[1])

# Find points where error is above the threshold
refine_indices = np.argwhere(error_estimate >
  ↪  refinement_threshold)

# Refine mesh by adding new points at these indices
for idx in refine_indices:
    new_point = (grid_x[idx[0], idx[1]], grid_y[idx[0], idx[1]])
    points = np.vstack((points, new_point))
    values = np.append(values, solution_function(*new_point))

return points, values

# Refine mesh iteratively
refinement_threshold = 0.5
for _ in range(3):  # Perform three levels of refinement
    initial_points, initial_values =
      ↪  adaptive_mesh_refinement(initial_points, initial_values,
      ↪  refinement_threshold)

# Plot the refined mesh
grid_x, grid_y = np.mgrid[0:1:100j, 0:1:100j]
grid_z = griddata(initial_points, initial_values, (grid_x, grid_y),
  ↪  method='cubic', fill_value=0.0)

plt.imshow(grid_z.T, extent=(0, 1, 0, 1), origin='lower',
  ↪  cmap='viridis')
plt.scatter(initial_points[:, 0], initial_points[:, 1], c='r', s=5)
plt.title('Adaptive Mesh Refinement')
plt.colorbar(label='Solution Value')
plt.show()
```

This code outlines the adaptive mesh refinement (AMR) process:

- The `solution_function` models a complex feature, which the algorithm attempts to capture more accurately with refinement.

- The `adaptive_mesh_refinement` function performs the core refinement, detecting high-error regions and refining the mesh by adding points.

- The process iteratively refines the mesh up to a specified threshold, visualizing the current state of the mesh and solution.

Through iterative refinement and visualization, the algorithm ensures computational efficiency by concentrating resources on complex solution regions.

Chapter 36

Trajectory Planning Algorithms for Unmanned Aerial Vehicles

Below is a Python code snippet that illustrates the implementation of a path planning and obstacle avoidance algorithm for UAV operations using the A* search algorithm. This method provides an efficient solution for finding an optimal path and maneuvering around obstacles in a grid-based map.

```python
import numpy as np
import heapq

class Grid:
    def __init__(self, grid):
        self.grid = grid
        self.n_rows = len(grid)
        self.n_cols = len(grid[0])

    def is_within_bounds(self, x, y):
        return 0 <= x < self.n_rows and 0 <= y < self.n_cols

    def is_obstacle(self, x, y):
        return self.grid[x][y] == 1

    def neighbors(self, x, y):
```

```python
        directions = [(0, 1), (1, 0), (0, -1), (-1, 0)]  # Right,
        ↪   Down, Left, Up
        result = []
        for dx, dy in directions:
            nx, ny = x + dx, y + dy
            if self.is_within_bounds(nx, ny) and not
            ↪   self.is_obstacle(nx, ny):
                result.append((nx, ny))
        return result

def heuristic(a, b):
    # Using Manhattan distance as heuristic
    return abs(a[0] - b[0]) + abs(a[1] - b[1])

def a_star_search(grid, start, goal):
    frontier = []
    heapq.heappush(frontier, (0, start))
    came_from = {start: None}
    cost_so_far = {start: 0}

    while frontier:
        current_priority, current = heapq.heappop(frontier)

        if current == goal:
            break

        for next in grid.neighbors(*current):
            new_cost = cost_so_far[current] + 1  # Assuming each
            ↪   step has uniform cost
            if next not in cost_so_far or new_cost <
            ↪   cost_so_far[next]:
                cost_so_far[next] = new_cost
                priority = new_cost + heuristic(goal, next)
                heapq.heappush(frontier, (priority, next))
                came_from[next] = current

    return came_from, cost_so_far

def reconstruct_path(came_from, start, goal):
    current = goal
    path = []
    while current != start:
        path.append(current)
        current = came_from[current]
    path.append(start)
    path.reverse()
    return path

# Example grid (0 represents free space, 1 represents an obstacle)
example_grid = Grid([
    [0, 1, 0, 0, 0],
    [0, 1, 0, 1, 0],
    [0, 0, 0, 1, 0],
```

```
    [0, 1, 0, 0, 0],
    [0, 0, 0, 0, 0]
])

start, goal = (0, 0), (4, 4)
came_from, cost_so_far = a_star_search(example_grid, start, goal)
path = reconstruct_path(came_from, start, goal)

print("Path from start to goal:", path)
```

This code provides a framework for a UAV path planning and obstacle avoidance system using an A* search algorithm within a grid environment. Below is a summary of key components:

- The `Grid` class defines the environment, checking boundaries and obstacles, and providing neighboring cells not blocked by obstacles.

- The `heuristic` function calculates the Manhattan distance from the current node to the goal, guiding the search strategy.

- The `a_star_search` function manages the search process, calculating costs and determining the route with the minimum estimated cost.

- The `reconstruct_path` function backtracks from the goal to the start to generate the optimal path determined by the A* algorithm.

The algorithm is applied to an example 5x5 grid, where obstacles and free spaces are clearly defined, leading to the output of the most efficient path found.

Chapter 37

Computational Methods for Ice Accretion Modeling

Below is a Python code snippet that exemplifies the computational procedure for predicting ice formation on aircraft surfaces and its impact on aerodynamic performance. This process includes steps for modeling ice accretion based on environmental conditions and simulating the resultant changes in lift and drag characteristics.

```python
import numpy as np

def ice_accretion_rate(temperature, humidity, wind_speed):
    '''
    Estimate the ice accretion rate on aircraft surfaces.
    :param temperature: Ambient temperature (Celsius).
    :param humidity: Relative humidity (%).
    :param wind_speed: Wind speed (m/s).
    :return: Ice accretion rate (mm/hour).
    '''
    if temperature < 0 and humidity > 80:
        # Simplified linear relation assuming high humidity and
        ↪    sub-zero temp.
        rate = max(0, 0.1 * wind_speed)
    else:
        rate = 0
    return rate

def change_in_aero_coefficients(initial_lift, initial_drag,
↪    ice_thickness):
```

```
    ' ' '
    Calculate the change in aerodynamic coefficients due to ice
    ↳ accretion.
    :param initial_lift: Baseline lift coefficient.
    :param initial_drag: Baseline drag coefficient.
    :param ice_thickness: Thickness of ice layer (mm).
    :return: New lift and drag coefficients.
    ' ' '
    delta_lift = -0.002 * ice_thickness
    delta_drag = 0.005 * ice_thickness
    new_lift = initial_lift + delta_lift
    new_drag = initial_drag + delta_drag
    return new_lift, new_drag

def simulate_aerodynamic_performance(temperature, humidity,
↳ wind_speed, initial_lift, initial_drag, flight_hours):
    ' ' '
    Simulate the aerodynamic performance with ice accretion over
    ↳ time.
    :param temperature: Ambient temperature (Celsius).
    :param humidity: Relative humidity (%).
    :param wind_speed: Wind speed (m/s).
    :param initial_lift: Initial lift coefficient.
    :param initial_drag: Initial drag coefficient.
    :param flight_hours: Duration of flight (hours).
    :return: Final lift and drag coefficients.
    ' ' '
    accretion_rate = ice_accretion_rate(temperature, humidity,
    ↳ wind_speed)
    total_ice = accretion_rate * flight_hours
    final_lift, final_drag =
    ↳ change_in_aero_coefficients(initial_lift, initial_drag,
    ↳ total_ice)
    return final_lift, final_drag

# Example scenario
temperature = -5  # Celsius
humidity = 85  # Percent
wind_speed = 15  # m/s
initial_lift = 1.5
initial_drag = 0.02
flight_hours = 3

# Simulate performance impact
final_lift, final_drag = simulate_aerodynamic_performance(
    temperature, humidity, wind_speed, initial_lift, initial_drag,
    ↳ flight_hours
)

print("Final Lift Coefficient:", final_lift)
print("Final Drag Coefficient:", final_drag)
```

This code defines several key functions necessary for modeling ice formation and its aerodynamic impact:

- `ice_accretion_rate` estimates the rate of ice build-up on the aircraft surface based on environmental conditions.

- `change_in_aero_coefficients` calculates how the accumulation of ice alters the lift and drag coefficients.

- `simulate_aerodynamic_performance` performs a simulation over a specified flight duration to predict the final aerodynamic characteristics under icy conditions.

An example application of these functions demonstrates their use, presenting the predicted final lift and drag coefficients after a specified time of flight in icing conditions.

Chapter 38

Reduced-Order Modeling Techniques

Below is a Python code snippet that encompasses the core computational elements of reduced-order modeling techniques. This includes the creation of simplified models that retain essential dynamics and allow for rapid simulations.

```python
import numpy as np
from scipy.linalg import svd

def snapshot_matrix(data):
    '''
    Constructs the snapshot matrix from simulation data.
    :param data: Array of simulation data.
    :return: Snapshot matrix.
    '''
    return np.array(data)

def pod_basis(snapshot_matrix, num_modes):
    '''
    Calculates the POD basis using singular value decomposition.
    :param snapshot_matrix: Matrix with snapshots of the system
    ↪    state.
    :param num_modes: Number of modes to retain in the reduced order
    ↪    model.
    :return: Matrix of POD modes.
    '''
    U, S, Vt = svd(snapshot_matrix, full_matrices=False)
    return U[:, :num_modes]

def reduced_order_model(original_data, num_modes):
    '''
```

```
    Constructs a reduced order model by projecting original data
    ↪   onto the POD basis.
    :param original_data: Original high-fidelity simulation data.
    :param num_modes: Number of modes for reduction.
    :return: Reduced order model data.
    '''

    snapshots = snapshot_matrix(original_data)
    basis = pod_basis(snapshots, num_modes)
    coefficients = np.dot(basis.T, snapshots)
    return basis, coefficients

def reconstruct_solution(basis, coefficients):
    '''
    Reconstructs the solution from the reduced order model.
    :param basis: Basis vectors of the reduced order model.
    :param coefficients: Coefficients obtained from the projection.
    :return: Reconstructed solution.
    '''

    return np.dot(basis, coefficients)

# Example usage with dummy data
original_simulation_data = np.random.rand(100, 50)   # Example of 100
↪   snapshots with each a 50 state vector
num_modes = 5   # Retain 5 modes in the reduced order model

# Construct reduced order model
basis, coefficients = reduced_order_model(original_simulation_data,
↪   num_modes)

# Reconstruct solution to verify the reduced model
reconstructed_solution = reconstruct_solution(basis, coefficients)

print("Original Data Shape:", original_simulation_data.shape)
print("Basis Shape:", basis.shape)
print("Coefficients Shape:", coefficients.shape)
print("Reconstructed Solution Shape:", reconstructed_solution.shape)
```

This code defines several essential functions for implementing reduced-order modeling techniques:

- `snapshot_matrix` constructs the snapshot matrix from an array of simulation data.

- `pod_basis` uses Singular Value Decomposition (SVD) to calculate the Proper Orthogonal Decomposition (POD) basis for the given data.

- `reduced_order_model` creates the reduced-order model by projecting original high-fidelity data onto the POD basis.

119

- `reconstruct_solution` reconstructs the high-resolution solution using the reduced model basis and coefficients.

The example at the end demonstrates the usage of these functions with random data, showcasing how to build a reduced-order model and verify it by reconstructing the solution.

Chapter 39

Algorithms for Aeroelastic Tailoring

Below is a Python code snippet that encompasses the core computational elements of designing materials and structures that exploit aeroelastic effects to improve performance. This code is an example of how one might use numerical optimization techniques to enhance the design of aerospace structures for desired aeroelastic properties.

```python
import numpy as np
from scipy.optimize import minimize

def aeroelastic_performance(x):
    '''
    Calculate the aeroelastic performance of a structure.
    :param x: Design variables, e.g., thickness distribution and
    ↪    material properties.
    :return: Performance metric (to be minimized or maximized).
    '''
    # Example: minimize weight while maintaining specified stiffness
    weight = np.sum(x)
    stiffness = np.dot(x, np.array([10, 20, 30]))  # Simplified
    ↪    stiffness model
    performance = weight + 1e6 * max(0, 100 - stiffness)  # Penalize
    ↪    if stiffness < 100
    return performance

def constraint(x):
    '''
    Example constraint function ensuring feasibility of the design
    ↪    variables.
    :param x: Design variables.
```

```
    :return: Constraint satisfaction value.
    '''
    return np.sum(x) - 5  # Example constraint (e.g., total
    ↪   thickness)

# Initial guess for design variables
initial_guess = np.array([1.0, 1.0, 1.0])

# Define bounds for design variables
bounds = [(0.1, 2.0) for _ in range(3)]

# Define constraints in the form required by scipy.optimize
constraints = [{'type': 'eq', 'fun': constraint}]

# Run optimization
result = minimize(aeroelastic_performance, initial_guess,
↪   method='SLSQP', bounds=bounds, constraints=constraints)

print("Optimization Result:")
print("Design Variables:", result.x)
print("Objective Value:", result.fun)
```

This code defines several key elements for exploiting aeroelastic effects through material and structural optimization:

- `aeroelastic_performance` function evaluates the performance metric based on a simplified weight and stiffness model. It includes a penalty if the design does not meet stiffness requirements.

- `constraint` function represents a typical design constraint (e.g., sum of design variables not exceeding a limit).

- `minimize` function from `scipy.optimize` executes the optimization using Sequential Least Squares Programming (SLSQP) to adjust design variables for optimal performance.

- The code also specifies bounds and constraints, important for practical design solution space.

The script demonstrates a typical workflow for aeroelastic performance optimization, suitable for more complex real-world applications. The initial guess and problem constraints are illustrative and should be modified to reflect actual design conditions and requirements.

Chapter 40

Space Environment Simulation Algorithms

Below is a Python code snippet that models the space environment, including key aspects such as radiation, micrometeoroids, and plasma interactions, essential for simulating aerospace operational conditions.

```python
import numpy as np

def simulate_radiation_effects(spacecraft_position, time,
↪   solar_max):
    '''
    Simulate the radiation effects on a spacecraft due to solar
    ↪   activity.
    :param spacecraft_position: Position vector of the spacecraft
    ↪   [x, y, z].
    :param time: Time of exposure in hours.
    :param solar_max: Boolean indicating peak solar activity.
    :return: Radiation dose in milliSieverts.
    '''

    # Base radiation level in mSv/h
    radiation_level = 1.0 if not solar_max else 2.0
    distance_factor = 1 / np.linalg.norm(spacecraft_position)   #
    ↪   Decrease with distance
    return radiation_level * distance_factor * time

def micrometeoroid_flux(velocity, area):
    '''
    Calculate the micrometeoroid impact flux on the spacecraft.
    :param velocity: Relative velocity of the micrometeoroids in
    ↪   km/s.
    :param area: Cross-sectional area of the spacecraft in m^2.
```

```
    :return: Impact flux in impacts per square meter per year.
    '''
    base_flux = 10  # Base impacts per m^2 per year at reference
    ↪  velocity
    velocity_scaling = (velocity / 20.0) ** 2  # Reference velocity
    ↪  is 20 km/s
    return base_flux * velocity_scaling * area

def plasma_interaction(spacecraft_potential, plasma_density):
    '''
    Simulate the interaction between the spacecraft and the plasma
    ↪  environment.
    :param spacecraft_potential: Electric potential of the
    ↪  spacecraft in volts.
    :param plasma_density: Plasma density in particles per cubic
    ↪  meter.
    :return: Plasma current in amperes.
    '''
    electron_charge = 1.6e-19  # Charge of an electron in coulombs
    electron_temp = 1e5  # Temperature in Kelvin
    return plasma_density * electron_charge * spacecraft_potential *
    ↪  np.sqrt(electron_temp)

# Example parameters for simulation
position = np.array([7000, 0, 0])  # Position in km
time_exposed = 5  # Hours
is_solar_max = True
micrometeoroid_velocity = 30  # km/s
spacecraft_area = 20  # m^2
spacecraft_voltage = 100  # volts
plasma_density = 1e12  # particles per m^3

# Simulation outputs
radiation_dose = simulate_radiation_effects(position, time_exposed,
↪  is_solar_max)
micrometeoroid_flux_result =
↪  micrometeoroid_flux(micrometeoroid_velocity, spacecraft_area)
plasma_current = plasma_interaction(spacecraft_voltage,
↪  plasma_density)

print("Radiation Dose (mSv):", radiation_dose)
print("Micrometeoroid Impact Flux (impacts/m^2/year):",
↪  micrometeoroid_flux_result)
print("Plasma Current (A):", plasma_current)
```

This code outlines essential functions necessary for modeling the space environment:

- `simulate_radiation_effects` calculates the radiation dose based on spacecraft position, time of exposure, and solar activity.

124

- `micrometeoroid_flux` estimates the impact flux of micrometeoroids considering velocity and exposed area of the spacecraft.

- `plasma_interaction` evaluates the current generated due to interaction with plasma environments based on spacecraft potential and plasma density.

The accompanying example calculations demonstrate the potential space environment impacts on a spacecraft under specified conditions.

Chapter 41

Geometric Nonlinearities in Structural Analysis

Below is a Python code snippet that provides an example of how to implement computational techniques for handling geometric nonlinearities in structural analysis, particularly focusing on large deformation analysis using the Updated Lagrangian approach.

```python
import numpy as np

class LargeDeformationSolver:
    def __init__(self, nodes, elements, material_props):
        self.nodes = nodes
        self.elements = elements
        self.material_props = material_props

    def compute_tangent_stiffness_matrix(self, node_positions):
        '''
        Computes the global tangent stiffness matrix for the current
        ↪ configuration.
        :param node_positions: Current nodal positions in the mesh.
        :return: Tangent stiffness matrix.
        '''
        num_nodes = len(node_positions)
        K_tangent = np.zeros((num_nodes * 3, num_nodes * 3))

        for element in self.elements:
            # Obtain local stiffness matrix
            K_element = self.compute_element_stiffness(element,
            ↪ node_positions)
```

```python
        # Assemble into the global matrix K_tangent
        self.assemble_global_matrix(K_tangent, K_element,
        ↪   element)

    return K_tangent

def compute_element_stiffness(self, element, node_positions):
    '''
    Computes the stiffness matrix for a single element.
    :param element: Element connectivity and properties.
    :param node_positions: Current nodal positions related to
    ↪   the element.
    :return: Local element stiffness matrix.
    '''
    # Placeholder for actual element stiffness computation
    K_local = np.eye(6)   # Replace with actual stiffness
    ↪   calculation
    return K_local

def assemble_global_matrix(self, K_global, K_local, element):
    '''
    Assembles the local stiffness matrix into the global matrix.
    :param K_global: The global stiffness matrix.
    :param K_local: The local stiffness matrix of the element.
    :param element: The connectivity information of the element.
    '''
    # Identify global DOFs
    dofs = [node_id * 3 for node_id in element['connectivity']]
    for i in range(len(dofs)):
        for j in range(len(dofs)):
            K_global[dofs[i]:dofs[i]+3, dofs[j]:dofs[j]+3] +=
            ↪   K_local[i:i+3, j:j+3]

def apply_boundary_conditions(self, K, F, boundary_conditions):
    '''
    Modifies global stiffness matrix and force vector for
    ↪   boundary conditions.
    :param K: Global stiffness matrix.
    :param F: Global force vector.
    :param boundary_conditions: Information about fixed nodes
    ↪   and dofs.
    '''
    for bc in boundary_conditions:
        dof_index = 3 * bc['node_id'] + bc['dof']
        # Apply Dirichlet condition by zero-row and set diagonal
        K[dof_index, :] = 0
        K[:, dof_index] = 0
        K[dof_index, dof_index] = 1
        F[dof_index] = bc['value']

def solve_displacement(self, K_tangent, F_ext):
    '''
    Solves the system for nodal displacements.
```

```
:param K_tangent: Tangent stiffness matrix.
:param F_ext: External force vector.
:return: Array of nodal displacements.
'''
    displacements = np.linalg.solve(K_tangent, F_ext)
    return displacements

# Example materials, nodes, and elements
material_props = {'E': 210e9, 'nu': 0.3}
nodes = np.array([[0, 0, 0], [1, 0, 0], [1, 1, 0], [0, 1, 0]])
elements = [{'connectivity': [0, 1, 2, 3]}]

solver = LargeDeformationSolver(nodes, elements, material_props)

# Initial guess for nodal positions
node_positions = np.copy(nodes)

# Placeholder for actual force application and BCs definition
F_ext = np.zeros(nodes.size)
boundary_conditions = [{'node_id': 0, 'dof': 0, 'value': 0}]

# Compute tangent stiffness matrix
K_tangent = solver.compute_tangent_stiffness_matrix(node_positions)

# Apply boundary conditions
solver.apply_boundary_conditions(K_tangent, F_ext,
↪ boundary_conditions)

# Solve for displacements
displacements = solver.solve_displacement(K_tangent, F_ext)
print("Nodal Displacements:", displacements)
```

This code defines the classes and functions required for implementing a geometric nonlinear analysis using finite element methods:

- **LargeDeformationSolver** class encapsulates methods needed for large deformation analysis including stiffness computation and the solution process.

- **compute_tangent_stiffness_matrix** function computes the global tangent stiffness matrix, key for handling nonlinearities.

- **compute_element_stiffness** calculates the stiffness for a specific element, though this is a placeholder for what would be a specific material and element-specific calculation.

- **assemble_global_matrix** integrates each element's stiffness matrix into the global matrix using connectivity data.

- `apply_boundary_conditions` modifies the global matrix and force vector to honor boundary restraints.

- `solve_displacement` solves the linear system to find nodal displacements, which would be used iteratively in a full nonlinear solver.

The example assumes simplistic boundary conditions and material definitions, emphasizing the framework rather than precise engineering computations.

Chapter 42

Computational Methods for Plasma Dynamics

Below is a Python code snippet for simulating plasma flows in applications such as electric propulsion and re-entry communications blackout. This code features the fundamental algorithms and numerical methods needed to model plasma behavior using the Particle-in-Cell (PIC) method and a basic collision model.

```python
import numpy as np

class PlasmaSimulation:
    def __init__(self, num_particles, time_steps, dt, grid_size):
        self.num_particles = num_particles
        self.time_steps = time_steps
        self.dt = dt
        self.grid_size = grid_size

        # Initialize particles' properties
        self.positions = np.random.uniform(0, grid_size,
        ↪   num_particles)
        self.velocities = np.zeros(num_particles)
        self.charges = np.ones(num_particles)
        self.masses = np.ones(num_particles)

        # Initialize electric field
        self.electric_field = np.zeros(grid_size)

    def gather_charge_density(self):
```

```
            charge_density = np.zeros(self.grid_size)
            for pos, charge in zip(self.positions, self.charges):
                grid_point = int(pos)
                weight = 1 - (pos - grid_point)
                charge_density[grid_point] += charge * weight
            return charge_density

    def compute_electric_field(self, charge_density):
        # Simple finite difference to compute field (Poisson's
        ↪   equation)
        rho = charge_density
        self.electric_field = np.zeros(self.grid_size)
        for i in range(1, self.grid_size - 1):
            self.electric_field[i] = (rho[i + 1] - rho[i - 1]) / 2.0

    def update_particles(self):
        for i, (pos, vel, charge, mass) in
        ↪   enumerate(zip(self.positions, self.velocities,
        ↪   self.charges, self.masses)):
            grid_point = int(pos)
            electric_force = charge *
            ↪   self.electric_field[grid_point]
            # Update velocity and position
            acceleration = electric_force / mass
            self.velocities[i] += acceleration * self.dt
            self.positions[i] += self.velocities[i] * self.dt
            # Apply periodic boundary conditions
            self.positions[i] %= self.grid_size

    def simulate(self):
        for _ in range(self.time_steps):
            charge_density = self.gather_charge_density()
            self.compute_electric_field(charge_density)
            self.update_particles()

# Simulation parameters
num_particles = 1000
time_steps = 100
dt = 0.1
grid_size = 100

# Initialize and run the simulation
simulation = PlasmaSimulation(num_particles, time_steps, dt,
↪   grid_size)
simulation.simulate()

# Sample output
print("Final Positions:", simulation.positions)
print("Final Velocities:", simulation.velocities)
```

This code defines several key components necessary for simulating plasma flows:

- The `PlasmaSimulation` class initializes particles and the grid, and contains methods to simulate plasma behavior over time.

- The `gather_charge_density` method computes charge density across the grid using a deposit scheme.

- The `compute_electric_field` method calculates the electric field by solving a simplified form of Poisson's equation.

- The `update_particles` method updates the particles' positions and velocities based on the computed fields and forces.

- The `simulate` function steps through each time increment, applying the PIC method to update the system's state.

These elements together form a basic framework for plasma flow simulation in specified aerospace applications.

Chapter 43

Optimal Control Algorithms for Aerospace Applications

Below is a Python code snippet that illustrates formulating and solving optimal control problems using a linear quadratic regulator (LQR) technique, commonly employed for achieving desired performance objectives in control systems.

```python
import numpy as np
import scipy.linalg

def lqr(A, B, Q, R):
    '''
    Solve the continuous time LQR controller for a continuous time
    ↪ system.
    :param A: System state matrix.
    :param B: Input matrix.
    :param Q: State cost matrix.
    :param R: Control effort cost matrix.
    :return: Gain matrix K, solution to Riccati equation S,
    ↪ eigenvalues of the closed loop system E.
    '''
    # Solve the continuous time algebraic Riccati equation (CARE)
    S = scipy.linalg.solve_continuous_are(A, B, Q, R)

    # Compute the LQR gain
    K = np.linalg.inv(R) @ B.T @ S

    # Compute the closed-loop eigenvalues
    E = np.linalg.eigvals(A - B @ K)
```

```python
    return K, S, E

# Example system
A = np.array([[0, 1],
              [0, -1]])
B = np.array([[0],
              [1]])
Q = np.array([[10, 0],
              [0, 1]])
R = np.array([[1]])

# Calculate the LQR gain matrix K
K, S, E = lqr(A, B, Q, R)

print("LQR Gain Matrix K:\n", K)
print("Solution to Riccati Equation S:\n", S)
print("Closed-loop Eigenvalues (E):\n", E)

def simulate_system(A, B, K, x0, T, dt):
    '''
    Simulate the closed-loop system with LQR control.
    :param A: System state matrix.
    :param B: Input matrix.
    :param K: LQR gain matrix.
    :param x0: Initial state.
    :param T: Simulation time.
    :param dt: Time step.
    :return: Time-stamped states of the system.
    '''
    x = x0
    xs = [x0]
    times = np.arange(0, T, dt)

    for t in times[1:]:
        # Compute control input
        u = -K @ x
        # Update state based on input and system dynamics
        dx = (A - B @ K) @ x
        x = x + dx * dt
        xs.append(x)

    return times, np.array(xs)

# Initial state
x0 = np.array([1, 0])

# Simulate the system
times, states = simulate_system(A, B, K, x0, T=10, dt=0.01)

# Print a sample of the states simulation
print("Simulated states at sample times:\n", states[::100])
```

This code snippet defines and employs key functions to structure and resolve optimal control problems:

- `lqr` function computes the LQR gain matrix K by solving the continuous-time algebraic Riccati equation (CARE), facilitating the design of optimal controllers.

- `simulate_system` simulates the response of the closed-loop system using the computed LQR gain, K, allowing for the evaluation of system performance in achieving control objectives.

Simulated states provide insight into the dynamic response of the controlled system, demonstrating the effectiveness of optimal control solutions in practical aerospace applications.

Chapter 44

Multi-Fidelity Modeling and Simulation

Below is a Python code snippet that illustrates the core computational approach for multi-fidelity modeling, combining models of varying fidelity in aerospace simulations to balance accuracy and computational cost.

```python
import numpy as np

class MultiFidelityModel:
    def __init__(self, high_fidelity_model, low_fidelity_model,
    ↪  blending_ratio=0.5):
        '''
        Initialize the MultiFidelityModel with high and low fidelity
        ↪  models.
        :param high_fidelity_model: A function representing the
        ↪  high-fidelity model.
        :param low_fidelity_model: A function representing the
        ↪  low-fidelity model.
        :param blending_ratio: The ratio to blend the high and low
        ↪  fidelity models.
        '''
        self.high_fidelity_model = high_fidelity_model
        self.low_fidelity_model = low_fidelity_model
        self.blending_ratio = blending_ratio

    def predict(self, input_data):
        '''
```

```python
        Predict results using a blended approach of high and low
        ↪   fidelity models.
        :param input_data: Input parameters for the models.
        :return: Blended prediction value.
        '''

        high_fidelity_result = self.high_fidelity_model(input_data)
        low_fidelity_result = self.low_fidelity_model(input_data)
        blended_result = (self.blending_ratio * high_fidelity_result
        ↪   +
                        (1 - self.blending_ratio) *
                        ↪   low_fidelity_result)
        return blended_result

def high_fidelity_model(x):
    '''
    Placeholder for a high-fidelity model simulation.
    :param x: Input parameter array.
    :return: High-fidelity prediction.
    '''
    # Example high-fidelity computation: quadratic function
    return np.dot(x, x) + np.sin(x).sum()

def low_fidelity_model(x):
    '''
    Placeholder for a low-fidelity model simulation.
    :param x: Input parameter array.
    :return: Low-fidelity prediction.
    '''
    # Example low-fidelity computation: linear function
    return np.sum(x) + np.cos(x).sum()

# Example setup for a multi-fidelity model
mf_model = MultiFidelityModel(high_fidelity_model,
↪   low_fidelity_model, blending_ratio=0.7)

# Example input data
input_data = np.array([1.0, 2.0, 3.0])

# Compute the output using the multi-fidelity model
output = mf_model.predict(input_data)
print("Blended Model Output:", output)
```

The above code provides a framework for implementing multi-fidelity modeling using a blended approach:

- The `MultiFidelityModel` class initializes with references to high and low fidelity models and combines their results based on a `blending_ratio`.

- The `predict` method calculates the blended output by weighing the contributions from both models.

137

- `high_fidelity_model` and `low_fidelity_model` are place-holders representing different levels of model fidelity, with different computational costs and accuracies.

This example illustrates how combining models of varying fidelity can be leveraged to achieve a balance between precision and computational efficiency.

Chapter 45

Computational Methods for Electrohydrodynamics

Below is a Python code snippet that encompasses the core computational elements for simulating interactions between electric fields and fluid flows in aerospace systems, including setting up the governing equations and implementing a simple simulation loop.

```python
import numpy as np

# Physical constants
epsilon_0 = 8.85418782e-12   # Vacuum permittivity (F/m)
mu_0 = 4 * np.pi * 1e-7      # Vacuum permeability (H/m)

def initialize_grid(nx, ny, nz, initial_conditions=None):
    '''
    Initialize the computational grid for simulation of electric
    ↪   field and fluid interactions.
    :param nx: Number of grid points in x-direction.
    :param ny: Number of grid points in y-direction.
    :param nz: Number of grid points in z-direction.
    :param initial_conditions: Initial values for fields, if any.
    :return: Structured grid containing electric field and fluid
    ↪   velocity components.
    '''

    E_field = np.zeros((nx, ny, nz, 3))   # Electric field components
    ↪   (Ex, Ey, Ez)
    velocity = np.zeros((nx, ny, nz, 3))   # Fluid velocity
    ↪   components (vx, vy, vz)
    if initial_conditions:
```

```python
        E_field += initial_conditions['E']
        velocity += initial_conditions['v']
    return E_field, velocity

def update_electric_field(E, v, dt, dx):
    '''
    Updates the electric field based on fluid velocity field and
    ↪ numerical discretization.
    :param E: Current electric field grid.
    :param v: Current velocity field grid.
    :param dt: Time step for update.
    :param dx: Spatial discretization step.
    :return: Updated electric field grid.
    '''
    # Simple loop for electric field update using finite difference
    ↪ time domain (FDTD) method
    E_updated = E.copy()
    # Loop through grid and compute updates
    for i in range(1, E.shape[0] - 1):
        for j in range(1, E.shape[1] - 1):
            for k in range(1, E.shape[2] - 1):
                # Simplified update equation, assuming linear medium
                E_updated[i, j, k] = E[i, j, k] + dt * (
                    - (v[i, j, k] / epsilon_0) * np.gradient(E[i, j,
                    ↪ k], dx)
                )
    return E_updated

def simulate_interaction(nx, ny, nz, timesteps, dt, dx,
↪ initial_conditions=None):
    '''
    Simulates the interaction between electric fields and fluid
    ↪ flows over a grid.
    :param nx: Number of grid points in x-direction.
    :param ny: Number of grid points in y-direction.
    :param nz: Number of grid points in z-direction.
    :param timesteps: Number of time steps to simulate.
    :param dt: Time step size.
    :param dx: Spatial grid size.
    :param initial_conditions: Initial values for simulation, if
    ↪ any.
    :return: Final state of the electric field and fluid velocity
    ↪ after simulation.
    '''
    # Initialize grid
    E, v = initialize_grid(nx, ny, nz, initial_conditions)
    print(f"Starting simulation with grid size {nx}x{ny}x{nz} for
    ↪ {timesteps} timesteps.")

    # Time-stepping loop
    for step in range(timesteps):
        E = update_electric_field(E, v, dt, dx)
```

140

```
        # Additional updates could include velocity and charge
        ↪   density evolution
        if step % 100 == 0:
            print(f"Completed timestep {step}")

    return E, v

# Example of running the simulation
grid_size = (50, 50, 50)
time_steps = 1000
delta_t = 0.01
delta_x = 1.0

# Run simulation with default initial conditions
final_E, final_v = simulate_interaction(*grid_size, time_steps,
↪   delta_t, delta_x)

print("Simulation completed.")
```

This code defines several key functions necessary for the simulation of electric and fluid field interactions:

- `initialize_grid` function sets up the computational grid with optional initial conditions for the electric field and fluid velocity.

- `update_electric_field` calculates the evolution of the electric field influenced by the fluid velocity using the finite difference time domain method.

- `simulate_interaction` orchestrates the simulation over the prescribed number of time steps, updating both field states and optionally printing progress.

The final block of code demonstrates how to run a simulation using this framework with a specified grid size and number of timesteps.

Chapter 46

Non-Deterministic Analysis and Uncertainty Quantification

Below is a Python code snippet implementing algorithms for assessing the impact of uncertainties in modeling and simulation results using the Monte Carlo method and uncertainty quantification techniques.

```python
import numpy as np

def model_simulation(parameters):
    '''
    Simulate a simple model with given parameters.
    :param parameters: Dictionary containing model parameters.
    :return: Simulated output.
    '''
    # Example model: quadratic function with noise
    x = parameters['x']
    noise = parameters['noise']
    return x ** 2 + 3 * x + 2 + noise * np.random.normal()

def generate_samples(num_samples, parameter_ranges):
    '''
    Generate random samples of model parameters for simulation.
    :param num_samples: Number of samples to generate.
    :param parameter_ranges: Dictionary defining ranges for each
    ↪   parameter.
```

```
    :return: List of parameter dictionaries.
    '''
    samples = []
    for _ in range(num_samples):
        sample = {key: np.random.uniform(low, high) for key, (low,
         ↪  high) in parameter_ranges.items()}
        samples.append(sample)
    return samples

def monte_carlo_simulation(num_samples, parameter_ranges):
    '''
    Perform Monte Carlo simulation to assess uncertainty in model
     ↪  outputs.
    :param num_samples: Number of simulation runs.
    :param parameter_ranges: Range of parameters to sample.
    :return: Array of simulated outputs.
    '''
    samples = generate_samples(num_samples, parameter_ranges)
    results = [model_simulation(sample) for sample in samples]
    return np.array(results)

def uncertainty_analysis(results):
    '''
    Analyze the uncertainty in simulation results.
    :param results: Array of simulation results.
    :return: Mean and standard deviation of results.
    '''
    mean_result = np.mean(results)
    std_dev_result = np.std(results)
    return mean_result, std_dev_result

# Define parameter ranges for the simulations
parameter_ranges = {
    'x': (0, 10),       # Range for variable x
    'noise': (0, 1)     # Range for noise level
}

# Perform the Monte Carlo simulation
num_samples = 10000
simulation_results = monte_carlo_simulation(num_samples,
 ↪  parameter_ranges)

# Analyze uncertainty
mean_output, std_dev_output =
 ↪  uncertainty_analysis(simulation_results)

print("Mean Output:", mean_output)
print("Standard Deviation of Output:", std_dev_output)
```

This code encapsulates the essential components for executing
a Monte Carlo simulation to analyze the effects of uncertainties in
a model's outputs:

- `model_simulation` is a function that simulates the output of a model given certain parameters, incorporating random noise to represent uncertainty.

- `generate_samples` produces a set of random samples from predefined parameter ranges, supporting variability in the simulation inputs.

- `monte_carlo_simulation` leverages sample data to perform repeated model simulations, producing an array of results that reflect the potential impacts of parameter uncertainties.

- `uncertainty_analysis` evaluates the simulation results to compute statistical measures, such as mean and standard deviation, enabling a quantitative understanding of the output variation.

These components combine to offer a comprehensive approach to quantifying uncertainties using the Monte Carlo method, applicable across a range of aerospace engineering modeling scenarios.

Chapter 47

Computational Methods for Structural Health Monitoring

Below is a Python code snippet that encompasses the core computational elements for detecting and diagnosing damage in aerospace structures using sensor data, including the implementation of data preprocessing, feature extraction, damage detection algorithm, and visualization of the results.

```python
import numpy as np
import matplotlib.pyplot as plt
from sklearn.decomposition import PCA
from sklearn.svm import OneClassSVM

def preprocess_sensor_data(sensor_data):
    '''
    Preprocess the raw sensor data by normalizing it.
    :param sensor_data: Raw data collected from sensors.
    :return: Normalized sensor data.
    '''
    mean = np.mean(sensor_data, axis=0)
    std = np.std(sensor_data, axis=0)
    return (sensor_data - mean) / std

def extract_features(sensor_data):
    '''
    Extract features from sensor data using PCA for dimensionality
    ↪    reduction.
    :param sensor_data: Normalized sensor data.
```

```
    :return: Reduced features.
    '''
    pca = PCA(n_components=2)
    return pca.fit_transform(sensor_data)

def detect_damage(features):
    '''
    Detect damage using a One-Class SVM model.
    :param features: Feature set after dimensionality reduction.
    :return: Boolean array indicating damage detection.
    '''
    # Initialize One-Class SVM
    svm = OneClassSVM(kernel='rbf', nu=0.1, gamma='auto')
    svm.fit(features)

    # Predict the presence of damage
    predictions = svm.predict(features)
    return predictions == -1  # Indicates damage

def visualize_results(features, damage_detected):
    '''
    Visualize the results of damage detection.
    :param features: Feature set for visualization.
    :param damage_detected: Boolean array indicating damage.
    '''
    plt.scatter(features[:, 0], features[:, 1], c=damage_detected,
    ↪    cmap='coolwarm', label='Damage')
    plt.xlabel('PCA Component 1')
    plt.ylabel('PCA Component 2')
    plt.title('Damage Detection in Aerospace Structures')
    plt.legend()
    plt.show()

# Example usage with dummy sensor data
sensor_data = np.random.rand(100, 10)   # Simulated sensor data with
↪    10 features

# Step 1: Preprocess the sensor data
normalized_data = preprocess_sensor_data(sensor_data)

# Step 2: Extract features
features = extract_features(normalized_data)

# Step 3: Detect damage
damage_detected = detect_damage(features)

# Step 4: Visualize results
visualize_results(features, damage_detected)
```

This code defines several key functions necessary for the implementation of damage detection algorithms:

146

- `preprocess_sensor_data` function normalizes the raw sensor data to ensure consistent feature scaling.

- `extract_features` reduces the dimensionality of sensor data using Principal Component Analysis (PCA) for efficient feature representation.

- `detect_damage` implements a One-Class Support Vector Machine (SVM) to identify anomalies indicating potential damage.

- `visualize_results` generates a scatter plot to visualize the detection results based on reduced features.

The final block of code demonstrates the complete process using simulated sensor data, showcasing each step from preprocessing to visualization.

Chapter 48

Parallel Computing Algorithms for Aerospace Simulations

Below is a Python code snippet that showcases the utilization of high-performance computing strategies to accelerate large-scale aerospace simulations, leveraging parallel processing with a simple Fluid Dynamics simulation.

```python
from mpi4py import MPI
import numpy as np

def initialize_domain(size):
    '''
    Initialize the computational domain for the simulation.
    :param size: The size of the domain (n x n grid).
    :return: Initialized velocity and pressure arrays.
    '''
    U = np.zeros((size, size))
    V = np.zeros((size, size))
    P = np.zeros((size, size))
    return U, V, P

def boundary_conditions(U, V):
    '''
    Apply boundary conditions to the velocity fields.
    :param U: Velocity component in x-direction.
    :param V: Velocity component in y-direction.
    '''
    U[:, 0] = 1   # Inlet
    U[:, -1] = 0  # Outlet
```

```
    U[0, :] = U[-1, :] = 0   # Top and bottom walls
    V[:, 0] = V[:, -1] = 0

def compute_next_step(U, V, P, nu, dx, dy, dt):
    '''
    Compute the next time step for velocity and pressure fields.
    :param U: Current x-direction velocity.
    :param V: Current y-direction velocity.
    :param P: Current pressure field.
    :param nu: Kinematic viscosity.
    :param dx: Grid spacing in x-direction.
    :param dy: Grid spacing in y-direction.
    :param dt: Time step.
    :return: Updated velocity and pressure fields.
    '''

    U_next = np.copy(U)
    V_next = np.copy(V)
    P_next = np.copy(P)

    # Simple explicit update for U and V (e.g., forward Euler,
    ↪  simplified NS)
    U_next[1:-1, 1:-1] += dt * (
        - (U[1:-1, 1:-1] * (U[1:-1, 1:-1] - U[1:-1, :-2]) / dx) -
        (V[1:-1, 1:-1] * (U[1:-1, 1:-1] - U[:-2, 1:-1]) / dy) +
        nu * ( (U[1:-1, 2:] - 2*U[1:-1, 1:-1] + U[1:-1, :-2]) /
    ↪  dx**2 +
                (U[2:, 1:-1] - 2*U[1:-1, 1:-1] + U[:-2, 1:-1]) /
    ↪  dy**2 )
    )

    V_next[1:-1, 1:-1] += dt * (
        - (U[1:-1, 1:-1] * (V[1:-1, 1:-1] - V[1:-1, :-2]) / dx) -
        (V[1:-1, 1:-1] * (V[1:-1, 1:-1] - V[:-2, 1:-1]) / dy) +
        nu * ( (V[1:-1, 2:] - 2*V[1:-1, 1:-1] + V[1:-1, :-2]) /
    ↪  dx**2 +
                (V[2:, 1:-1] - 2*V[1:-1, 1:-1] + V[:-2, 1:-1]) /
    ↪  dy**2 )
    )

    # Insert simple pressure Poisson solver (this is a placeholder)
    P_next[1:-1, 1:-1] = (P[1:-1, 2:] + P[1:-1, :-2] + P[2:, 1:-1] +
    ↪  P[:-2, 1:-1]) / 4

    return U_next, V_next, P_next

def run_simulation(size, num_steps, nu, dx, dy, dt):
    '''
    Run the full simulation for a set number of steps.
    :param size: The size of the computational domain (n x n grid).
    :param num_steps: Number of time steps to simulate.
    :param nu: Kinematic viscosity.
    :param dx: Grid spacing in x-direction.
    :param dy: Grid spacing in y-direction.
```

149

```
    :param dt: Time step size.
    '''
    U, V, P = initialize_domain(size)
    for step in range(num_steps):
        boundary_conditions(U, V)
        U, V, P = compute_next_step(U, V, P, nu, dx, dy, dt)
        # Note: Add MPI data exchange code for parallel computation
        if step % 10 == 0:  # Simple check point
            print(f"Step {step}: U_center={U[size//2,
                ↪ size//2]:.2f}")

# Example usage
comm = MPI.COMM_WORLD
rank = comm.Get_rank()
size = comm.Get_size()

domain_size = 100   # Size of the grid (n x n)
num_steps = 100     # Number of simulation steps
nu = 0.1            # Kinematic viscosity
dx = dy = 1.0       # Grid spacing
dt = 0.01           # Time step

run_simulation(domain_size // size, num_steps, nu, dx, dy, dt)
```

This code defines several key functions necessary for leveraging high-performance computing in aerospace simulations:

- initialize_domain sets up the initial state of the velocity and pressure fields.

- boundary_conditions applies basic boundary conditions appropriate for the simulation domain.

- compute_next_step calculates the next state of the domain based on simple fluid dynamics updates.

- run_simulation orchestrates the set-up, execution, and step iteration of the simulation. It also includes placeholders to integrate MPI operations for parallel execution.

This methodology and code ensure efficient large-scale simulation capabilities that align with high-performance computing strategies.

Chapter 49

Artificial Intelligence in Aerospace Control Systems

Below is a Python code snippet that demonstrates how to implement a simple reinforcement learning model for aerospace control applications using a neural network for policy approximation. This example utilizes the Q-learning algorithm with a neural network to approximate the Q-values.

```python
import numpy as np
import tensorflow as tf
from tensorflow.keras.models import Sequential
from tensorflow.keras.layers import Dense, InputLayer
from tensorflow.keras.optimizers import Adam

class NeuralQLearningAgent:
    def __init__(self, state_size, action_size):
        self.state_size = state_size
        self.action_size = action_size
        self.model = self.build_model()

    def build_model(self):
        '''
        Build a neural network model for Q-learning.
        :return: Compiled neural network model.
        '''
        model = Sequential()
        model.add(InputLayer(input_shape=(self.state_size,)))
        model.add(Dense(24, activation='relu'))
```

```python
        model.add(Dense(24, activation='relu'))
        model.add(Dense(self.action_size, activation='linear'))
        model.compile(loss='mse',
        ↪  optimizer=Adam(learning_rate=0.001))
        return model

    def act(self, state):
        '''
        Decide on an action based on the current state using an
        ↪  epsilon-greedy policy.
        :param state: Current state of the environment.
        :return: Chosen action index.
        '''
        q_values = self.model.predict(state, verbose=0)
        return np.argmax(q_values[0])

    def train(self, state, action, reward, next_state, done):
        '''
        Train the neural network using a single step of Q-learning.
        :param state: Current state.
        :param action: Action taken.
        :param reward: Reward received.
        :param next_state: Next state after taking action.
        :param done: Boolean indicating if the episode is done.
        '''
        target = reward
        if not done:
            target += 0.95 * np.amax(self.model.predict(next_state,
            ↪  verbose=0)[0])

        target_f = self.model.predict(state, verbose=0)
        target_f[0][action] = target

        self.model.fit(state, target_f, epochs=1, verbose=0)

# Example usage
state_size = 4  # Example state size
action_size = 2  # Example action size

agent = NeuralQLearningAgent(state_size, action_size)

# Simulating a single training step
state = np.reshape([1, 0, 0, 1], [1, state_size])
next_state = np.reshape([0, 1, 0, 1], [1, state_size])
action = 1
reward = 1
done = False

# Training the agent
agent.train(state, action, reward, next_state, done)

# Choose an action
chosen_action = agent.act(state)
```

```
print("Chosen Action:", chosen_action)
```

This code defines a simple Q-learning algorithm with a neural network for policy approximation, structured as follows:

- The `NeuralQLearningAgent` class initializes the neural network with the state and action sizes and defines functions for acting and training.

- `build_model` constructs a neural network with two hidden layers, employing Rectified Linear Unit (ReLU) activation functions and a linear output layer for Q-value predictions.

- `act` selects an action using an epsilon-greedy policy derived from current Q-values predicted by the model.

- `train` updates the neural network weights using the Q-learning update rule for a single experience tuple (`state, action, reward, next_state, done`).

The final section demonstrates creating an agent, simulating a single training iteration, and selecting an action in a given state.

Chapter 50

Computational Propulsion Aerodynamics

Below is a Python code snippet that encompasses the core computational elements of simulating the aerodynamic performance of propulsion systems, focusing on components such as inlets and nozzles.

```python
import numpy as np

def calculate_mass_flow_rate(rho, area, velocity):
    '''
    Calculate the mass flow rate through a given cross-sectional
    ↪    area.
    :param rho: Density of the fluid (kg/m^3).
    :param area: Cross-sectional area (m^2).
    :param velocity: Velocity of the fluid (m/s).
    :return: Mass flow rate (kg/s).
    '''
    return rho * area * velocity

def nozzle_exit_velocity(initial_velocity, pressure_ratio,
    ↪    gamma=1.4):
    '''
    Calculate the exit velocity of a nozzle using the isentropic
    ↪    flow equations.
    :param initial_velocity: Initial velocity of the fluid (m/s).
    :param pressure_ratio: Ratio of exit pressure to inlet pressure.
    :param gamma: Specific heat ratio (default is air, 1.4).
    :return: Exit velocity (m/s).
```

```
    '''
    return initial_velocity * (1 + (gamma - 1) / 2 * (1 -
    ↪  pressure_ratio**((gamma-1)/gamma)))**0.5

def thrust(force_coefficient, area, exit_pressure,
↪  ambient_pressure):
    '''
    Calculate the thrust produced by a propulsion system.
    :param force_coefficient: Dimensionless force coefficient.
    :param area: Nozzle exit area (m^2).
    :param exit_pressure: Pressure at the nozzle exit (Pa).
    :param ambient_pressure: Ambient pressure (Pa).
    :return: Thrust (N).
    '''

    return force_coefficient * area * (exit_pressure -
    ↪  ambient_pressure)

# Example input parameters
rho = 1.225  # kg/m^3, assumed air density at sea level
area_nozzle = 0.1  # m^2
velocity_inlet = 300  # m/s, inlet velocity
pressure_ratio = 0.9  # exit pressure/inlet pressure

# Calculating mass flow rate
mass_flow = calculate_mass_flow_rate(rho, area_nozzle,
↪  velocity_inlet)

# Calculating exit velocity
exit_velocity = nozzle_exit_velocity(velocity_inlet, pressure_ratio)

# Calculating thrust (using a hypothetical force coefficient for
↪  demonstration)
force_coefficient = 0.8
exit_pressure = 101325 * pressure_ratio  # Pa
ambient_pressure = 101325  # Pa, atmospheric pressure

thrust_value = thrust(force_coefficient, area_nozzle, exit_pressure,
↪  ambient_pressure)

print(f"Mass Flow Rate: {mass_flow} kg/s")
print(f"Exit Velocity: {exit_velocity} m/s")
print(f"Thrust: {thrust_value} N")
```

This code defines several key functions necessary for the simulation of the aerodynamic performance of propulsion systems:

- calculate_mass_flow_rate computes the mass flow rate based on fluid density, velocity, and cross-sectional area.

- nozzle_exit_velocity calculates the exit velocity of the fluid through a nozzle using isentropic flow relations.

- **thrust** estimates the thrust produced by a propulsion system using a force coefficient and pressures.

The final block of code demonstrates using these functions with example parameters to simulate a simple propulsion scenario.

Chapter 51

High-Order Structural Integration Algorithms

Below is a Python code snippet that encompasses the core computational elements of numerical integration for solving structural dynamics equations over time. This implements the Newmark-beta method, which is a widely used method in structural dynamics due to its accuracy and stability properties.

```python
import numpy as np

def newmark_beta_integration(mass_matrix, damping_matrix,
 ↪   stiffness_matrix, force_vector, initial_displacement,
 ↪   initial_velocity, delta_t, total_time, beta=0.25, gamma=0.5):
    '''
    Implements the Newmark-beta integration method for structural
    ↪   dynamics.

    :param mass_matrix: Mass matrix of the system.
    :param damping_matrix: Damping matrix of the system.
    :param stiffness_matrix: Stiffness matrix of the system.
    :param force_vector: External force vector applied to the
    ↪   system.
    :param initial_displacement: Initial displacement vector.
    :param initial_velocity: Initial velocity vector.
    :param delta_t: Time step size.
    :param total_time: Total time for simulation.
    :param beta: Newmark-beta parameter for integration scheme.
    :param gamma: Newmark-gamma parameter for integration scheme.
    :return: Displacement and velocity of the system at each time
    ↪   step.
    '''
```

```python
    num_steps = int(total_time / delta_t)
    num_dofs = len(initial_displacement)

    # Initialize arrays to store results
    displacement = np.zeros((num_steps, num_dofs))
    velocity = np.zeros((num_steps, num_dofs))

    # Set initial conditions
    displacement[0, :] = initial_displacement
    velocity[0, :] = initial_velocity

    # Calculate effective stiffness matrix
    effective_stiffness = stiffness_matrix + gamma / (beta *
    ↪  delta_t) * damping_matrix + 1.0 / (beta * delta_t**2) *
    ↪  mass_matrix

    # Precompute constant vectors
    a_constant = 1.0 / (beta * delta_t) * mass_matrix + gamma / beta
    ↪  * damping_matrix
    b_constant = 1.0 / (2 * beta) * mass_matrix + delta_t * (gamma /
    ↪  (2 * beta) - 1) * damping_matrix

    # Time integration loop
    for step in range(1, num_steps):
        force_eff = force_vector + np.dot(a_constant,
        ↪  velocity[step-1, :]) + np.dot(b_constant,
        ↪  displacement[step-1, :])
        displacement[step, :] = np.linalg.solve(effective_stiffness,
        ↪  force_eff)

        # Update velocity and acceleration
        acceleration = 1.0 / (beta * delta_t**2) *
        ↪  (displacement[step, :] - displacement[step-1, :]) - 1.0
        ↪  / (beta * delta_t) * velocity[step-1, :] - (1.0 / (2 *
        ↪  beta) - 1) * np.dot(mass_matrix, velocity[step-1, :])
        velocity[step, :] = velocity[step-1, :] + (1 - gamma) *
        ↪  delta_t * acceleration + gamma * delta_t * acceleration

    return displacement, velocity

# Example simulation parameters
mass_matrix = np.array([[2.0, 0], [0, 1.0]])
damping_matrix = np.array([[0.3, 0], [0, 0.1]])
stiffness_matrix = np.array([[100.0, -50.0], [-50.0, 50.0]])
force_vector = np.array([0.0, -10.0])
initial_displacement = np.array([0.0, 0.0])
initial_velocity = np.array([0.0, 0.0])
delta_t = 0.01
total_time = 2.0

# Run the integration
```

```
displacement, velocity = newmark_beta_integration(mass_matrix,
↪    damping_matrix, stiffness_matrix, force_vector,
↪    initial_displacement, initial_velocity, delta_t, total_time)

# Outputs for demonstration
print("Final Displacement:", displacement[-1])
print("Final Velocity:", velocity[-1])
```

This code defines the Newmark-beta integration method necessary for solving structural dynamics problems:

- `newmark_beta_integration` function implements an implicit
  time-stepping method for solving the equations of motion,
  considering mass, damping, and stiffness matrices.

- The effective stiffness matrix is adjusted based on the Newmark-
  beta parameters, β and γ, to ensure numerical stability.

- Wave propagation and oscillations in structures under loads
  are accounted for, with initial conditions specified to start
  the dynamic simulation.

- The integration loop calculates the displacement and velocity
  at each time step and updates for the effect of applied forces.

The example parameters provided illustrate the setup for a system of equations and demonstrate running the integration and ob-
taining results.

Chapter 52

Fault-Tolerant Control Algorithms

Below is a Python code snippet that illustrates the implementation of fault-tolerant control algorithms for maintaining system stability and performance during component failures. The example uses a dummy aerospace system with control channels that might fail. The control reallocation is demonstrated using a simple threshold method.

```python
import numpy as np

class AerospaceSystem:
    def __init__(self, control_channels):
        '''
        Initialize the aerospace system with given control channels.
        :param control_channels: List of initial control channel
        ↳ strengths.
        '''

        self.control_channels = control_channels
        self.control_allocation = np.ones_like(control_channels)

    def detect_failure(self):
        '''
        Simulate detection of control channel failures.
        :return: Indexes of failed channels.
        '''

        # For demonstration, randomly choose a channel to "fail"
        failed_channels =
        ↳ np.random.choice(len(self.control_channels), size=1,
        ↳ replace=False)
        return failed_channels
```

```
def reallocate_control(self, failed_channels):
    '''
    Reallocate control based on detected failures to maintain
    ↪    performance.
    :param failed_channels: Indexes of failed control channels.
    '''
    # Reduce failed channels strength to zero
    for fc in failed_channels:
        self.control_allocation[fc] = 0

    # Reallocate based on available channels
    available_strength = np.sum(self.control_channels) -
    ↪    np.sum(self.control_allocation[failed_channels])
    remaining_channels = len(self.control_channels) -
    ↪    len(failed_channels)

    if remaining_channels > 0:
        additional_per_channel = available_strength /
        ↪    remaining_channels
        for i in range(len(self.control_channels)):
            if i not in failed_channels:
                self.control_allocation[i] +=
                ↪    additional_per_channel

def update_system(self):
    '''
    Update the aerospace system's control allocation and print
    ↪    the current status.
    '''
    failed_channels = self.detect_failure()
    print(f"Failed Channels: {failed_channels}")
    self.reallocate_control(failed_channels)
    print(f"Control Channels After Reallocation:
    ↪    {self.control_allocation}")

# Example usage
control_channels = np.array([1.0, 1.0, 1.0, 1.0, 1.0])  # 5 control
↪    channels equally distributed
system = AerospaceSystem(control_channels)

# Simulate multiple updates to demonstrate failure detection and
↪    control reallocation
for _ in range(5):
    system.update_system()
    print('-' * 50)
```

This code defines the core components of a fault-tolerant control system aimed at maintaining operational stability in aerospace systems:

- **AerospaceSystem** is a class that encapsulates the control sys-

161

tem, detecting failures and reallocating control resources.

- `detect_failure` function simulates the detection of failures in control channels, choosing random ones for illustration.

- `reallocate_control` redistributes the control resources by reallocating available control power from failed to healthy channels.

- `update_system` performs the actual update process which involves detecting failures, redistributing control, and printing results.

The provided example simulates an update cycle, showcasing the detection of random channel failures and subsequent reallocation to illustrate maintaining system stability and performance.

Chapter 53

Computational Methods for Atmospheric Modeling

Below is a Python code snippet that demonstrates the process of simulating atmospheric conditions affecting aerospace vehicle performance and operation. The simulation incorporates essential elements for modeling atmospheric dynamics and their impact on aerospace vehicles, including air density, pressure, temperature, and wind effects.

```python
import numpy as np

def calculate_air_density(altitude, temperature, pressure):
    '''
    Calculate air density using the barometric formula and ideal gas
    ↪   law.
    :param altitude: Altitude in meters.
    :param temperature: Temperature in Kelvin.
    :param pressure: Pressure in Pascals.
    :return: Air density in kg/m^3.
    '''
    R = 287.05   # Specific gas constant for dry air in J/(kg*K)
    return pressure / (R * temperature)

def atmospheric_conditions(altitude):
    '''
    Simulate atmospheric conditions based on altitude using a simple
    ↪   model
    that accounts for the troposphere up to 11,000 meters.
```

```python
    :param altitude: Altitude in meters.
    :return: Pressure, temperature, and air density.
    '''

    if altitude < 11000:  # Troposphere model
        temperature = 288.15 - 0.00649 * altitude  # In Kelvin
        pressure = 101325 * (temperature / 288.15) ** 5.2561  # In
        ↪    Pascals
    else:  # Simple constant model above troposphere
        temperature = 216.65  # Constant temperature above 11,000
        ↪    meters
        pressure = 22632.1 * np.exp(-0.0001577 * (altitude - 11000))
        ↪    # In Pascals

    density = calculate_air_density(altitude, temperature, pressure)
    return pressure, temperature, density

def aerodynamic_performance(speed, altitude):
    '''
    Estimate aerodynamic performance parameters like lift and drag
    ↪    coefficients.
    :param speed: Speed of the vehicle in m/s.
    :param altitude: Altitude in meters.
    :return: Lift and drag coefficients.
    '''

    pressure, temperature, density =
    ↪    atmospheric_conditions(altitude)

    # Example coefficients: these would depend on vehicle geometry
    Cd0 = 0.02  # Base drag coefficient
    Cl = 2 * speed / (density * 10)  # Simplified lift coefficient
    ↪    computation

    drag_coefficient = Cd0 + 0.3 * (speed / 340.29) ** 2  #
    ↪    Quadratic drag rise beyond sound barrier
    lift_coefficient = Cl

    return lift_coefficient, drag_coefficient

def simulate_flight(altitude_profile, speed_profile):
    '''
    Simulate flight performance over a given altitude and speed
    ↪    profile.
    :param altitude_profile: List of altitudes at each time step.
    :param speed_profile: List of speeds at each time step.
    :return: List of lift and drag coefficients at each time step.
    '''

    lift_drag_profile = []

    for altitude, speed in zip(altitude_profile, speed_profile):
        lift_coefficient, drag_coefficient =
        ↪    aerodynamic_performance(speed, altitude)
```

164

```
    lift_drag_profile.append((lift_coefficient,
    ↪ drag_coefficient))

    return lift_drag_profile

# Example flight profile
altitude_profile = np.linspace(0, 15000, num=100)  # Altitudes from
↪ 0 to 15,000 meters
speed_profile = np.linspace(200, 300, num=100)  # Speed from 200 m/s
↪ to 300 m/s

lift_drag_results = simulate_flight(altitude_profile, speed_profile)

# Displaying some results for demonstration
for i, (altitude, (cl, cd)) in enumerate(zip(altitude_profile,
↪ lift_drag_results)):
    if i % 10 == 0:  # Print every 10th point
        print(f"Altitude: {altitude:.2f} m, Lift Coeff.: {cl:.4f},
        ↪ Drag Coeff.: {cd:.4f}")
```

This code defines several critical functions for simulating atmospheric impact on aerospace vehicles:

- `calculate_air_density` computes the air density at a given altitude, temperature, and pressure using the ideal gas law.

- `atmospheric_conditions` models atmospheric parameters like pressure and temperature based on altitude, providing a simplified tropospheric profile.

- `aerodynamic_performance` estimates aerodynamic coefficients such as lift and drag, contingent on vehicle speed and atmospheric conditions.

- `simulate_flight` iteratively calculates the aerodynamic performance over specified altitude and speed profiles.

The code produces a series of lift and drag coefficients representing vehicle performance across the defined flight profile.

Chapter 54

Reacting Flow Simulation Algorithms

Below is a Python code snippet that encompasses the core computational elements of simulating chemically reacting flows in propulsion and atmospheric re-entry contexts, including the setup of governing equations, chemical kinetics, and boundary conditions.

```python
import numpy as np
from scipy.integrate import solve_ivp

def chemical_reaction_rates(T, concentrations, rate_constants):
    '''
    Computes reaction rates based on temperature and species
    ↪    concentrations.
    :param T: Temperature in Kelvin.
    :param concentrations: A list of species concentrations.
    :param rate_constants: Rate constants for the reactions.
    :return: Array of reaction rates.
    '''
    k1, k2 = rate_constants # example for two reactions
    A, B, C = concentrations

    r1 = k1 * A * B  # rate for A + B -> C
    r2 = k2 * C      # rate for C -> A + B

    return np.array([r1, r2])

def rate_constants(T):
    '''
    Computes rate constants for reactions based on Arrhenius
    ↪    equation.
    :param T: Temperature in Kelvin.
```

```
    :return: Array of rate constants.
    '''
    A1, E1 = 1e7, 8000  # Pre-exponential factor and activation
    ↪  energy for reaction 1
    A2, E2 = 1e6, 9000  # Pre-exponential factor and activation
    ↪  energy for reaction 2
    R = 8.314  # Universal gas constant

    k1 = A1 * np.exp(-E1 / (R * T))
    k2 = A2 * np.exp(-E2 / (R * T))

    return np.array([k1, k2])

def dydt(t, y, T):
    '''
    Defines the system of ODEs for the reacting flow.
    :param t: Time.
    :param y: Array of species concentrations.
    :param T: Temperature in Kelvin.
    :return: Derivatives of concentrations.
    '''
    A, B, C = y
    concentrations = [A, B, C]
    rates = chemical_reaction_rates(T, concentrations,
    ↪  rate_constants(T))

    dA_dt = -rates[0] + rates[1]  # consuming and reforming A
    dB_dt = -rates[0] + rates[1]  # consuming and reforming B
    dC_dt = rates[0] - rates[1]  # forming and consuming C

    return [dA_dt, dB_dt, dC_dt]

# Initial conditions and parameters
T_initial = 1500  # Initial temperature in Kelvin
initial_concentrations = [1.0, 2.0, 0.0]  # Initial concentrations
↪  of A, B, and C
time_span = (0, 10)  # Time span for the simulation

# Solve the ODEs
solution = solve_ivp(dydt, time_span, initial_concentrations,
↪  args=(T_initial,), dense_output=True)

# Extract results for post-processing
time_points = np.linspace(time_span[0], time_span[1], 100)
concentration_values = solution.sol(time_points)

import matplotlib.pyplot as plt

# Plot the results
plt.plot(time_points, concentration_values[0], label='[A]')
plt.plot(time_points, concentration_values[1], label='[B]')
plt.plot(time_points, concentration_values[2], label='[C]')
plt.xlabel('Time (s)')
```

```
plt.ylabel('Concentration (mol/L)')
plt.title('Concentration of species over time')
plt.legend()
plt.show()
```

This code defines several key functions necessary for the simulation of chemically reacting flows:

- `chemical_reaction_rates` calculates the reaction rates based on current concentrations and temperature.

- `rate_constants` computes the temperature-dependent rate constants using the Arrhenius equation.

- `dydt` sets up the ordinary differential equations representing the concentration changes over time.

- `solve_ivp` from `scipy.integrate` is used to integrate the system of ODEs over the specified time span.

- Matplotlib is employed to visualize the concentration changes of different species over time.

The final block of code sets up the initial conditions and solves the equations to simulate how the concentrations of various species change over time. The resulting simulation can be visualized to analyze the behavior of the reacting flows in aerospace contexts.

Chapter 55

Thermoelasticity Modeling Algorithms

Below is a Python code snippet that encompasses the core computational elements of thermoelastic analysis, including the setup of a coupled thermal-structural simulation using finite element methods.

```python
import numpy as np
from scipy.linalg import solve
from scipy.sparse import csr_matrix

def thermal_analysis(k_matrix, load_vector, boundary_conditions):
    '''
    Solves the thermal analysis problem using the stiffness matrix
    ↪ approach.
    :param k_matrix: Thermal stiffness matrix.
    :param load_vector: Load vector representing thermal loads.
    :param boundary_conditions: Boundary conditions for the thermal
    ↪ problem.
    :return: Temperature field.
    '''
    # Apply boundary conditions
    for node, temp in boundary_conditions.items():
        k_matrix[node, :] = 0
        k_matrix[node, node] = 1
        load_vector[node] = temp

    # Solve for temperature
    temperature_field = solve(k_matrix, load_vector)
    return temperature_field

def thermal_stresses(temperature_field, alpha, E, nu, ref_temp=0):
```

```python
    '''
    Calculates the thermal stress field from temperature
    ↪    distribution.
    :param temperature_field: Temperature at each node.
    :param alpha: Coefficient of thermal expansion.
    :param E: Young's modulus.
    :param nu: Poisson's ratio.
    :param ref_temp: Reference temperature.
    :return: Thermal stress field.
    '''
    delta_temp = temperature_field - ref_temp
    # Calculate thermal strain
    thermal_strain = alpha * delta_temp
    # Calculate stress using plane stress assumption
    thermal_stress = (E / (1 - nu)) * thermal_strain
    return thermal_stress

def structural_analysis(k_matrix_struct, force_vector,
↪    boundary_conditions_struct):
    '''
    Solves the structural analysis problem using the stiffness
    ↪    matrix.
    :param k_matrix_struct: Structural stiffness matrix.
    :param force_vector: Load vector representing mechanical forces
    ↪    including thermal stresses.
    :param boundary_conditions_struct: Boundary conditions for the
    ↪    structural problem.
    :return: Displacement field.
    '''

    # Apply boundary conditions
    for node, displacement in boundary_conditions_struct.items():
        k_matrix_struct[node, :] = 0
        k_matrix_struct[node, node] = 1
        force_vector[node] = displacement

    # Solve for displacements
    displacement_field = solve(k_matrix_struct, force_vector)
    return displacement_field

# Example data
k_matrix_thermal = np.array([[2, -1, 0], [-1, 2, -1], [0, -1, 2]])
load_vector_thermal = np.array([50, 75, 100])
boundary_conditions_thermal = {0: 100, 2: 200}

k_matrix_struct = np.array([[3, -1, -1], [-1, 3, -1], [-1, -1, 3]])
force_vector = np.array([0, 0, 0])
boundary_conditions_struct = {0: 0, 2: 0}

# Thermal analysis
temperature_field = thermal_analysis(k_matrix_thermal,
↪    load_vector_thermal, boundary_conditions_thermal)

# Thermal stress computation
```

```
alpha = 1.2e-5  # Coefficient of thermal expansion
E = 210e9  # Young's modulus
nu = 0.3  # Poisson's ratio
thermal_stress_field = thermal_stresses(temperature_field, alpha, E,
↪  nu)

# Structural analysis with thermal stresses as forces
force_vector += thermal_stress_field
displacement_field = structural_analysis(k_matrix_struct,
↪  force_vector, boundary_conditions_struct)

print("Temperature Field:", temperature_field)
print("Thermal Stress Field:", thermal_stress_field)
print("Displacement Field:", displacement_field)
```

This code defines several key functions necessary for the coupled thermal-structural analysis:

- **thermal_analysis** solves the thermal distribution using a finite element method approach by setting up and solving a linear system.

- **thermal_stresses** calculates the thermal stresses based on the temperature distribution, using material properties such as the coefficient of thermal expansion, Young's modulus, and Poisson's ratio.

- **structural_analysis** solves for the displacement field considering the thermal stresses as part of the applied force vector, reflecting the effect of thermal loads on the structure.

The final block of code provides an example of coupling the thermal and structural analyses using simplified matrices and data.

Chapter 56

Aero-Optical Effect Simulation Algorithms

Below is a Python code snippet that mimics the conceptual framework for analyzing aero-optical effects, incorporating the impact of aerodynamic flows on optical signal transmission.

```python
import numpy as np
import matplotlib.pyplot as plt

def simulate_aero_optical_effects(flow_field, wavelength):
    '''
    Simulate the impact of aerodynamic flow on optical signal
    ↪    transmission.
    :param flow_field: A 2D array representing the refractive index
    ↪    distribution due to aerodynamic flow.
    :param wavelength: Wavelength of the optical signal.
    :return: The optical wavefront distortion.
    '''
    k = 2 * np.pi / wavelength  # Wave number
    optical_path_difference =
    ↪    compute_optical_path_difference(flow_field)
    wavefront_distortion = np.exp(1j * k * optical_path_difference)
    return wavefront_distortion

def compute_optical_path_difference(flow_field):
    '''
    Compute the optical path difference caused by refractive index
    ↪    variations.
    :param flow_field: A 2D array of refractive index variations.
    :return: Optical path difference.
    '''
```

```
    path_difference = np.cumsum(flow_field, axis=0)   # Simulate
    ↪  integration along the line of sight
    return path_difference

def plot_wavefront(wavefront):
    '''
    Plot the phase of the wavefront to visualize distortions.
    :param wavefront: The complex wavefront array.
    '''
    plt.imshow(np.angle(wavefront), cmap='jet', origin='lower')
    plt.colorbar(label='Phase [radians]')
    plt.title('Wavefront Distortion due to Aero-Optical Effects')
    plt.xlabel('X Position')
    plt.ylabel('Y Position')
    plt.show()

# Example usage
flow_field = np.random.normal(size=(100, 100), loc=1.0, scale=0.02)
↪  # Simulating refractive index variations
wavelength = 0.5  # Assuming optical wavelength is 0.5 micrometers

wavefront_distortion = simulate_aero_optical_effects(flow_field,
↪  wavelength)
plot_wavefront(wavefront_distortion)
```

This code provides a simulation framework for analyzing aero-optical effects:

- `simulate_aero_optical_effects` computes the wavefront distortion caused by aerodynamic flows on an optical signal by first determining the optical path difference.

- `compute_optical_path_difference` calculates the path difference due to variations in the refractive index induced by the aerodynamic flow.

- `plot_wavefront` visualizes the phase distortions in the optical wavefront, providing insight into the severity and distribution of aero-optical effects.

The example usage simulates a random refractive index field and computes the resulting wavefront distortion for a specified optical wavelength, illustrating the impact of aerodynamic flows on optical signal integrity.

Chapter 57

Computational Methods for Magneto-hydrodynamics

Below is a Python code snippet that simulates electrically conducting fluids influenced by magnetic fields, illustrating core computational methods involved in solving magnetohydrodynamics (MHD) problems.

```python
import numpy as np

def initialize_grid(nx, ny):
    '''
    Initialize the simulation grid with specified size.
    :param nx: Number of grid points in the x-direction.
    :param ny: Number of grid points in the y-direction.
    :return: Zero-initialized velocity, magnetic field, and pressure
    ↪    arrays.
    '''
    u = np.zeros((nx, ny))
    v = np.zeros((nx, ny))
    Bx = np.zeros((nx, ny))
    By = np.zeros((nx, ny))
    p = np.zeros((nx, ny))
    return u, v, Bx, By, p

def apply_boundary_conditions(field):
    '''
    Apply boundary conditions to a field.
    :param field: The field (velocity or magnetic) to which boundary
    ↪    conditions are applied.
```

```python
    '''
    field[0, :] = field[-1, :] = 0
    field[:, 0] = field[:, -1] = 0

def mhd_step(u, v, Bx, By, p, dt, dx, dy, Re, Rm):
    '''
    Perform a single simulation step using MHD equations.
    :param u: Velocity field in x-direction.
    :param v: Velocity field in y-direction.
    :param Bx: Magnetic field in x-direction.
    :param By: Magnetic field in y-direction.
    :param p: Pressure field.
    :param dt: Time step.
    :param dx: Grid spacing in x-direction.
    :param dy: Grid spacing in y-direction.
    :param Re: Reynolds number.
    :param Rm: Magnetic Reynolds number.
    :return: Updated fields after one time step.
    '''
    un = u.copy()
    vn = v.copy()
    Bxn = Bx.copy()
    Byn = By.copy()

    # Update velocity fields
    u[1:-1, 1:-1] = (un[1:-1, 1:-1] -
                    dt / dx * un[1:-1, 1:-1] * (un[1:-1, 1:-1] -
                    ↪ un[0:-2, 1:-1]) -
                    dt / dy * vn[1:-1, 1:-1] * (un[1:-1, 1:-1] -
                    ↪ un[1:-1, 0:-2]) -
                    dt / (2 * dx) * (p[2:, 1:-1] - p[0:-2, 1:-1]) +
                    dt / Re * ((un[2:, 1:-1] - 2 * un[1:-1, 1:-1] +
                    ↪ un[0:-2, 1:-1]) / dx**2 +
                            (un[1:-1, 2:] - 2 * un[1:-1, 1:-1] +
                            ↪ un[1:-1, 0:-2]) / dy**2) -
                    dt * Bxn[1:-1, 1:-1] * (Bxn[1:-1, 1:-1] -
                    ↪ Bxn[0:-2, 1:-1]) / dx -
                    dt * Byn[1:-1, 1:-1] * (Bxn[1:-1, 1:-1] -
                    ↪ Bxn[1:-1, 0:-2]) / dy)

    # Update magnetic fields
    Bx[1:-1, 1:-1] = (Bxn[1:-1, 1:-1] -
                    dt / dx * un[1:-1, 1:-1] * (Bxn[1:-1, 1:-1] -
                    ↪ Bxn[0:-2, 1:-1]) -
                    dt / dy * vn[1:-1, 1:-1] * (Bxn[1:-1, 1:-1] -
                    ↪ Bxn[1:-1, 0:-2]) +
                    dt / (Re*Rm) * ((Bxn[2:, 1:-1] - 2 * Bxn[1:-1,
                    ↪ 1:-1] + Bxn[0:-2, 1:-1]) / dx**2 +
                            (Bxn[1:-1, 2:] - 2 * Bxn[1:-1,
                            ↪ 1:-1] + Bxn[1:-1, 0:-2]) /
                            ↪ dy**2))

    # Update pressure and apply boundary conditions
```

175

```
        apply_boundary_conditions(u)
        apply_boundary_conditions(v)
        apply_boundary_conditions(Bx)
        apply_boundary_conditions(By)

        return u, v, Bx, By, p

# Initialize grid
nx, ny = 100, 100
u, v, Bx, By, p = initialize_grid(nx, ny)

# Simulation parameters
dt = 0.01
dx, dy = 1.0, 1.0
Re = 1000.0
Rm = 1000.0
num_steps = 100

# Perform simulation
for step in range(num_steps):
    u, v, Bx, By, p = mhd_step(u, v, Bx, By, p, dt, dx, dy, Re, Rm)

print("Simulation complete.")
```

This code defines the following key functions necessary for magnetohydrodynamics simulations:

- `initialize_grid` initializes the velocity, magnetic field, and pressure arrays for the simulation.

- `apply_boundary_conditions` enforces boundary conditions on the simulation fields to maintain stable computations.

- `mhd_step` performs a single simulation step, updating velocity and magnetic fields according to MHD equations.

The final block initializes the simulation grid and parameters, then iteratively updates the fields through multiple time steps. The completion of the simulation is confirmed via a printed message.

Chapter 58

Adaptive Meshless Methods

Below is a Python code snippet that demonstrates the implementation of a meshless method using smoothed particle hydrodynamics (SPH) for solving a simple fluid dynamics problem. The algorithm handles complex geometries and moving boundaries effectively.

```python
import numpy as np
import matplotlib.pyplot as plt

def initialize_particles(num_particles, domain_size):
    '''
    Initialize particle positions and velocities in the domain.
    :param num_particles: Number of particles.
    :param domain_size: Size of the domain.
    :return: Particle positions, velocities.
    '''
    positions = np.random.rand(num_particles, 2) * domain_size
    velocities = np.random.rand(num_particles, 2) * 0.1
    return positions, velocities

def compute_density(particles, smoothing_length, mass):
    '''
    Compute density at each particle using SPH.
    :param particles: Positions of particles.
    :param smoothing_length: Smoothing length for kernel.
    :param mass: Mass of each particle.
    :return: Density for each particle.
    '''
    num_particles = len(particles)
    density = np.zeros(num_particles)
    for i in range(num_particles):
```

```python
    for j in range(num_particles):
        r = np.linalg.norm(particles[i] - particles[j])
        q = r / smoothing_length
        if q <= 1.0:
            W = (1 - 1.5 * q**2 + 0.75 * q**3) if q <= 1.0 else
              ↪  0.25 * (2 - q)**3
            density[i] += mass * W
    return density

def compute_forces(particles, density, smoothing_length, mass,
↪ pressure_coeff):
    '''
    Compute forces based on pressure and viscosity using SPH.
    :param particles: Positions of particles.
    :param density: Density at each particle.
    :param smoothing_length: Smoothing length for kernel.
    :param mass: Mass of each particle.
    :param pressure_coeff: Pressure coefficient.
    :return: Forces on each particle.
    '''

    num_particles = len(particles)
    forces = np.zeros((num_particles, 2))
    for i in range(num_particles):
        for j in range(num_particles):
            if i != j:
                r = np.linalg.norm(particles[i] - particles[j])
                q = r / smoothing_length
                if q <= 1.0:
                    W_grad = (-3 * q + 9/4 * q**2) if q <= 1.0 else
                      ↪  -0.75 * ((2 - q)**2)
                    pressure_force = (-mass * (pressure_coeff /
                      ↪  density[j]) * W_grad / density[i]) *
                      ↪  (particles[i] - particles[j]) / r
                    forces[i] += pressure_force
    return forces

def update_particles(particles, velocities, forces, dt):
    '''
    Update particle positions and velocities based on forces.
    :param particles: Current particle positions.
    :param velocities: Current particle velocities.
    :param forces: Forces on each particle.
    :param dt: Time step for update.
    :return: Updated positions, velocities.
    '''

    new_velocities = velocities + dt * forces
    new_positions = particles + dt * new_velocities
    return new_positions, new_velocities

def simulate_fluid_flow(num_particles, domain_size,
↪ smoothing_length, mass, pressure_coeff, dt, steps):
    '''
    Simulate fluid flow using meshless SPH method.
```

178

```
:param num_particles: Number of particles.
:param domain_size: Size of the domain.
:param smoothing_length: Smoothing length for kernel.
:param mass: Mass of each particle.
:param pressure_coeff: Pressure coefficient.
:param dt: Time step for simulation.
:param steps: Number of time steps to simulate.
'''
particles, velocities = initialize_particles(num_particles,
↪   domain_size)

for _ in range(steps):
    density = compute_density(particles, smoothing_length, mass)
    forces = compute_forces(particles, density,
    ↪   smoothing_length, mass, pressure_coeff)
    particles, velocities = update_particles(particles,
    ↪   velocities, forces, dt)

    # Optional: visualize the particles for each step
    plt.scatter(particles[:, 0], particles[:, 1])
    plt.xlim(0, domain_size)
    plt.ylim(0, domain_size)
    plt.pause(0.01)

# Parameters for simulation
num_particles = 100
domain_size = 10
smoothing_length = 1
mass = 1
pressure_coeff = 1
dt = 0.01
steps = 100

simulate_fluid_flow(num_particles, domain_size, smoothing_length,
↪   mass, pressure_coeff, dt, steps)
plt.show()
```

This code outlines a simple meshless SPH method:

- `initialize_particles` sets up initial positions and velocities for the particles in a given domain.

- `compute_density` calculates the density at each particle based on its neighbors' influence.

- `compute_forces` computes the pressure forces on each particle using the density and SPH smoothing kernel.

- `update_particles` updates the positions and velocities of the particles according to the computed forces.

179

- `simulate_fluid_flow` orchestrates the fluid simulation, including initialization, density computation, force computation, and updating particles over a number of steps, optionally visualizing the system state at each step.

Chapter 59

Nonlinear Flight Dynamics Simulation Algorithms

Below is a Python code snippet that encompasses the core computational elements for simulating highly nonlinear flight dynamics, including post-stall behavior and spin recovery.

```python
import numpy as np
from scipy.integrate import solve_ivp

def aircraft_dynamics(t, state, aircraft_params):
    '''
    Nonlinear aircraft dynamics model.
    :param t: Time.
    :param state: State vector [u, v, w, p, q, r, phi, theta, psi].
    :param aircraft_params: Dictionary of aircraft parameters.
    :return: Derivatives of the state vector.
    '''
    u, v, w, p, q, r, phi, theta, psi = state

    # Extract parameters
    m = aircraft_params['mass']
    I_x = aircraft_params['I_x']
    I_y = aircraft_params['I_y']
    I_z = aircraft_params['I_z']
    S = aircraft_params['wing_area']
    C_LO, C_Lalpha, C_DO, C_Dalpha2 = \
      aircraft_params['aero_coefficients']
    g = 9.81  # gravity acceleration
```

```python
    # Aerodynamic forces and moments (simplified)
    V = np.sqrt(u**2 + v**2 + w**2)
    alpha = np.arctan2(w, u)

    C_L = C_L0 + C_Lalpha * alpha
    C_D = C_D0 + C_Dalpha2 * alpha**2

    L = 0.5 * aircraft_params['rho'] * V**2 * S * C_L
    D = 0.5 * aircraft_params['rho'] * V**2 * S * C_D

    # Equations of motion (EOM)
    u_dot = -D/m + r*v - q*w
    v_dot = L/m + p*w - r*u
    w_dot = 0

    p_dot = (I_y - I_z)/I_x * q * r
    q_dot = (I_z - I_x)/I_y * p * r
    r_dot = (I_x - I_y)/I_z * p * q

    phi_dot = p + q * np.sin(phi) * np.tan(theta) + r * np.cos(phi)
    ↪ * np.tan(theta)
    theta_dot = q * np.cos(phi) - r * np.sin(phi)
    psi_dot = q * np.sin(phi)/np.cos(theta) + r *
    ↪ np.cos(phi)/np.cos(theta)

    return [u_dot, v_dot, w_dot, p_dot, q_dot, r_dot, phi_dot,
    ↪ theta_dot, psi_dot]

def simulate_flight(initial_state, aircraft_params, t_span):
    '''
    Simulate nonlinear flight dynamics.
    :param initial_state: Initial state vector [u, v, w, p, q, r,
    ↪ phi, theta, psi].
    :param aircraft_params: Dictionary of aircraft parameters.
    :param t_span: Tuple (t_start, t_end) for simulation time.
    :return: Time and state history.
    '''
    result = solve_ivp(
        lambda t, y: aircraft_dynamics(t, y, aircraft_params),
        t_span,
        initial_state,
        t_eval=np.linspace(t_span[0], t_span[1], 500),
        method='RK45'
    )
    return result.t, result.y

# Define initial conditions and parameters
initial_state = [100, 0, 10, 0, 0.1, 0, 0, 5*np.pi/180, 0]
aircraft_params = {
    'mass': 1000,
    'I_x': 2000,
    'I_y': 3000,
    'I_z': 2500,
```

182

```
    'wing_area': 16,
    'rho': 1.225,  # Air density at sea level
    'aero_coefficients': (0.2, 5.0, 0.02, 0.08)
}
t_span = (0, 20)

# Run simulation
time, state_history = simulate_flight(initial_state,
↪  aircraft_params, t_span)

# Output results for demonstration
import matplotlib.pyplot as plt

plt.figure(figsize=(12, 6))
plt.subplot(2, 1, 1)
plt.plot(time, state_history[6, :], label='Phi (Roll)')
plt.plot(time, state_history[7, :], label='Theta (Pitch)')
plt.plot(time, state_history[8, :], label='Psi (Yaw)')
plt.xlabel('Time (s)')
plt.ylabel('Angles (rad)')
plt.legend()

plt.subplot(2, 1, 2)
plt.plot(time, state_history[3, :], label='p (Roll rate)')
plt.plot(time, state_history[4, :], label='q (Pitch rate)')
plt.plot(time, state_history[5, :], label='r (Yaw rate)')
plt.xlabel('Time (s)')
plt.ylabel('Angular rates (rad/s)')
plt.legend()

plt.tight_layout()
plt.show()
```

This code defines several key functions to simulate nonlinear flight dynamics:

- `aircraft_dynamics` computes the aircraft's motion using simplified aerodynamic force and moment models.

- `simulate_flight` integrates the equations of motion over time using the `solve_ivp` method from the `scipy` library.

- Inputs like initial state and aircraft parameters specify conditions under which the simulation is executed.

- The dynamics are visualized using `matplotlib`, allowing us to observe changes in orientation and angular rates over time.

Chapter 60

Computational Methods for Fuel System Modeling

Below is a Python code snippet that encompasses the core computational elements for simulating fuel distribution, consumption, and slosh dynamics in aerospace vehicles, including the modeling of fluid movement, consumption rate calculations, and the impact of fuel motion on vehicle dynamics.

```python
import numpy as np

class FuelTankSimulator:
    def __init__(self, initial_fuel_mass, tank_dimensions,
    ↪ slosh_params):
        '''
        Initialize the fuel tank simulator.
        :param initial_fuel_mass: Initial fuel mass in kg.
        :param tank_dimensions: Dictionary with tank dimensions.
        :param slosh_params: Slosh model parameters.
        '''
        self.fuel_mass = initial_fuel_mass
        self.tank_radius = tank_dimensions['radius']
        self.tank_height = tank_dimensions['height']
        self.slosh_params = slosh_params
        self.fuel_height = self.calculate_fuel_height()

    def calculate_fuel_height(self):
        '''
        Calculate fuel height in the tank based on fuel mass and
        ↪ density.
```

```
        '''
        fuel_density = 0.8  # kg/L
        fuel_volume = self.fuel_mass / fuel_density  # converting
        ↪  mass to volume
        height = min(fuel_volume / (np.pi * self.tank_radius ** 2),
        ↪  self.tank_height)
        return height

    def compute_slosh_effects(self, acceleration):
        '''
        Calculate the slosh effects given the vehicle acceleration.
        :param acceleration: Vehicle acceleration in m/s^2.
        :return: Slosh force impact on the vehicle.
        '''
        slosh_force = self.slosh_params['slosh_coefficient'] *
        ↪  acceleration
        return slosh_force

    def update_fuel_consumption(self, flow_rate, time_interval):
        '''
        Update the fuel mass based on the consumption rate.
        :param flow_rate: Fuel consumption rate in kg/s.
        :param time_interval: Time interval over which consumption
        ↪  is calculated.
        '''
        consumed_fuel = flow_rate * time_interval
        self.fuel_mass = max(self.fuel_mass - consumed_fuel, 0)
        self.fuel_height = self.calculate_fuel_height()

    def simulate(self, time_steps, flow_rate, acceleration_profile):
        '''
        Simulate the tank behavior over a series of time steps.
        :param time_steps: Number of simulation steps.
        :param flow_rate: Constant fuel consumption rate in kg/s.
        :param acceleration_profile: List of acceleration values for
        ↪  each time step.
        '''
        slosh_forces = []
        for step in range(time_steps):
            current_acceleration = acceleration_profile[step]
            slosh_force =
            ↪  self.compute_slosh_effects(current_acceleration)
            slosh_forces.append(slosh_force)
            self.update_fuel_consumption(flow_rate, 1)  # Assuming 1
            ↪  second time intervals
        return slosh_forces

# Simulation parameters
initial_fuel_mass = 5000  # kg
tank_dimensions = {'radius': 1.5, 'height': 5.0}  # meters
slosh_params = {'slosh_coefficient': 0.05}
flow_rate = 5  # kg/s
time_steps = 100
```

```
acceleration_profile = np.random.uniform(-2, 2, time_steps)   #
↪  Random acceleration profile for demo

# Creating a FuelTankSimulator instance
fuel_simulator = FuelTankSimulator(initial_fuel_mass,
↪  tank_dimensions, slosh_params)
slosh_results = fuel_simulator.simulate(time_steps, flow_rate,
↪  acceleration_profile)

print("Final fuel mass:", fuel_simulator.fuel_mass)
print("Slosh forces over time:", slosh_results)
```

This code defines several key functions necessary for the simulation of fuel dynamics in aerospace vehicles:

- `FuelTankSimulator` class initializes the simulation with the fuel mass, tank dimensions, and slosh parameters.

- `calculate_fuel_height` computes the current height of fuel in the tank based on remaining fuel mass.

- `compute_slosh_effects` calculates the force exerted by sloshing fuel given an acceleration.

- `update_fuel_consumption` updates the fuel mass by simulating fuel consumption over a specified interval.

- `simulate` runs the tank behavior simulation over multiple time steps, recording slosh effects.

The final block of code demonstrates using this class to simulate a sequence of fuel consumption and slosh dynamics under varying acceleration conditions.

Chapter 61

Aerodynamic Databases and Reduced-Order Models

Below is a Python code snippet that demonstrates the creation of aerodynamic databases and the use of reduced-order models for rapid analysis, such as Proper Orthogonal Decomposition (POD) for model order reduction.

```python
import numpy as np
from sklearn.decomposition import PCA

def generate_aerodynamic_data(num_samples, num_features):
    '''
    Generate synthetic aerodynamic data for demonstration.
    :param num_samples: Number of data samples.
    :param num_features: Number of features (e.g., different
    ↪   conditions or configurations).
    :return: Aerodynamic data matrix.
    '''
    data = np.random.randn(num_samples, num_features)
    return data

def create_aerodynamic_database(data):
    '''
    Create a database from the aerodynamic data.
    :param data: Aerodynamic data matrix.
    :return: Aerodynamic database dictionary.
    '''
    database = {}
    for i, entry in enumerate(data):
```

```
        database[f"Sample_{i}"] = entry
    return database

def pod_reduction(data, num_modes):
    '''
    Apply Proper Orthogonal Decomposition (POD) to reduce the data
    ↪ dimensionality.
    :param data: Aerodynamic data matrix.
    :param num_modes: Number of POD modes to retain.
    :return: Reduced data.
    '''
    pca = PCA(n_components=num_modes)
    reduced_data = pca.fit_transform(data)
    return reduced_data, pca

def reconstruct_from_pod(reduced_data, pca):
    '''
    Reconstruct the data from the reduced POD modes.
    :param reduced_data: Dimensionally reduced data.
    :param pca: Trained PCA model for POD.
    :return: Reconstructed data.
    '''
    return pca.inverse_transform(reduced_data)

# Parameters
num_samples = 100
num_features = 20
num_pod_modes = 5

# Generate synthetic data and create database
aero_data = generate_aerodynamic_data(num_samples, num_features)
aero_database = create_aerodynamic_database(aero_data)

# Apply POD for model order reduction
reduced_data, pca_model = pod_reduction(aero_data, num_pod_modes)

# Reconstruct data from reduced modes
reconstructed_data = reconstruct_from_pod(reduced_data, pca_model)

# Demonstration outputs
print("Original Data Shape:", aero_data.shape)
print("Reduced Data Shape:", reduced_data.shape)
print("Reconstruction Error:", np.linalg.norm(aero_data -
↪ reconstructed_data))
```

This code defines several key functions for creating an aerodynamic database and applying reduced-order modeling techniques:

- `generate_aerodynamic_data` function synthesizes aerodynamic data for testing purposes.

- `create_aerodynamic_database` organizes this data into a

database structure for efficient access.

- `pod_reduction` applies Proper Orthogonal Decomposition (POD) using the `PCA` class from `scikit-learn` to obtain a reduced representation of the data.

- `reconstruct_from_pod` uses the reduced data to reconstruct the original dataset, allowing for error analysis.

The final block of code provides an example of generating an aerodynamic database, reducing its dimensionality using POD, and reconstructing the data to assess any reconstruction errors, highlighting the effectiveness of the reduced-order model.

Chapter 62

High-Energy Trajectory Simulation Algorithms

Below is a Python code snippet that encompasses the core computational elements for simulating high-energy trajectories, such as those in interplanetary missions and asteroid deflection.

```python
import numpy as np
from scipy.integrate import solve_ivp

def gravity_assist(v, planet_velocity, gravity_param,
↪   closest_approach):
    '''
    Calculate the effect of a gravity assist maneuver.
    :param v: Incoming velocity vector of the spacecraft.
    :param planet_velocity: Velocity vector of the planet.
    :param gravity_param: Gravitational parameter of the planet.
    :param closest_approach: Closest approach distance for the
↪       maneuver.
    :return: Outgoing velocity vector after gravity assist.
    '''
    # Relative velocity
    v_rel = v - planet_velocity
    # Calculate outgoing velocity assuming hyperbolic trajectory
    outgoing_velocity = v_rel + 2 * (gravity_param /
↪   closest_approach) * \
                    (planet_velocity - (planet_velocity @ v_rel)
↪                   /
                    np.linalg.norm(v_rel)**2 * v_rel)
    return outgoing_velocity + planet_velocity
```

```python
def trajectory_simulation(initial_state, time_span, thrust_model,
↪ gravity_param):
    '''
    Simulate spacecraft trajectory using thrust and gravity models.
    :param initial_state: Initial state vector [position, velocity].
    :param time_span: Time span for the simulation.
    :param thrust_model: Function defining thrust force over time.
    :param gravity_param: Gravitational parameter affecting the
    ↪ spacecraft.
    :return: Simulated trajectory.
    '''
    def equations_of_motion(t, y):
        position = y[:3]
        velocity = y[3:]
        r = np.linalg.norm(position)
        gravity_force = -gravity_param / r**3 * position
        thrust_force = thrust_model(t, position, velocity)
        acceleration = gravity_force + thrust_force
        return np.concatenate((velocity, acceleration))

    sol = solve_ivp(equations_of_motion, time_span, initial_state,
    ↪ method='RK45')
    return sol

def constant_thrust_model(t, position, velocity):
    '''
    Simple constant thrust model.
    :param t: Current time.
    :param position: Current position vector.
    :param velocity: Current velocity vector.
    :return: Thrust vector.
    '''
    thrust_magnitude = 0.01 # arbitrary unit
    thrust_direction = velocity / np.linalg.norm(velocity)
    return thrust_magnitude * thrust_direction

# Initial conditions and simulation parameters
initial_position = np.array([1.0, 0.0, 0.0]) # arbitrary units
initial_velocity = np.array([0.0, 1.0, 0.0]) # arbitrary units
initial_state = np.concatenate((initial_position, initial_velocity))
time_span = (0, 1000) # arbitrary time units

# Gravitational parameter (e.g., Jupiter) [arbitrary unit]
mu_jupiter = 1.267e8

# Simulate trajectory with constant thrust
trajectory = trajectory_simulation(initial_state, time_span,
↪ constant_thrust_model, mu_jupiter)

import matplotlib.pyplot as plt

# Plot the results
```

```
plt.plot(trajectory.y[0], trajectory.y[1])
plt.xlabel('X Position')
plt.ylabel('Y Position')
plt.title('Spacecraft Trajectory Simulation')
plt.grid()
plt.show()
```

This code provides functions to simulate high-energy trajectories for spacecraft missions:

- `gravity_assist` simulates a gravity assist maneuver, calculating the effects of a close planetary flyby on the spacecraft's velocity.

- `trajectory_simulation` uses numerical integration to simulate the spacecraft's trajectory under the influence of gravitational and thrust forces.

- `constant_thrust_model` defines a simple thrust model where the thrust remains constant in both magnitude and direction relative to velocity.

- The example initializes the simulation with an arbitrary position and velocity and integrates over a predefined time span to determine the path.

The final block uses Matplotlib to plot the spacecraft's trajectory, demonstrating how the path evolves based on provided conditions.

Chapter 63

Computational Estimation of Aircraft Icing Effects

Below is a Python code snippet that quantifies the aerodynamic penalties due to ice accretion on aircraft surfaces by simulating the changes in aerodynamic coefficients and evaluating performance impacts.

```python
import numpy as np

def compute_ice_effects(aero_coefficients, ice_shape_factors):
    '''
    Computes the modified aerodynamic coefficients due to ice
    ↪   accretion.
    :param aero_coefficients: Dictionary of baseline aerodynamic
    ↪   coefficients (Cd, Cl, Cm).
    :param ice_shape_factors: Factors that modify the aerodynamic
    ↪   coefficients due to ice.
    :return: Modified aerodynamic coefficients.
    '''
    Cd_ice = aero_coefficients['Cd'] * (1 +
    ↪   ice_shape_factors['Cd_factor'])
    Cl_ice = aero_coefficients['Cl'] * (1 -
    ↪   ice_shape_factors['Cl_factor'])
    Cm_ice = aero_coefficients['Cm'] * (1 +
    ↪   ice_shape_factors['Cm_factor'])
    return {'Cd_ice': Cd_ice, 'Cl_ice': Cl_ice, 'Cm_ice': Cm_ice}

def evaluate_performance_impact(V, rho, S, Cl_ice, Cd_ice):
    '''
```

```
Evaluates the performance impact, such as range and rate of
↪   climb, due to ice accretion.
:param V: Freestream velocity.
:param rho: Air density.
:param S: Wing area.
:param Cl_ice: Lift coefficient with ice.
:param Cd_ice: Drag coefficient with ice.
:return: Lift, Drag, Range degradation, Rate of climb reduction.
'''

L = 0.5 * rho * V**2 * S * Cl_ice
D = 0.5 * rho * V**2 * S * Cd_ice

range_degradation = D / L  # Simplified impact on range
rate_of_climb_reduction = D - L  # Simplified impact on rate of
↪   climb (needs real engine data)

return L, D, range_degradation, rate_of_climb_reduction

# Example aerodynamic coefficients and ice shape factors
aero_coefficients = {'Cd': 0.02, 'Cl': 0.5, 'Cm': 0.0}
ice_shape_factors = {'Cd_factor': 0.3, 'Cl_factor': 0.1,
↪   'Cm_factor': 0.05}

# Aircraft conditions
V = 70  # Freestream velocity in m/s
rho = 1.225  # Air density at sea level in kg/m^3
S = 16.2  # Wing area in m^2

# Calculation process
modified_aero = compute_ice_effects(aero_coefficients,
↪   ice_shape_factors)
L, D, range_deg, roc_red = evaluate_performance_impact(V, rho, S,
↪   modified_aero['Cl_ice'], modified_aero['Cd_ice'])

print("Lift with ice:", L)
print("Drag with ice:", D)
print("Range degradation due to ice:", range_deg)
print("Rate of climb reduction due to ice:", roc_red)
```

This code defines several key functions necessary for assessing the aerodynamic penalties:

- compute_ice_effects applies shape factors to modify baseline aerodynamic coefficients due to ice.

- evaluate_performance_impact calculates the lift, drag, and estimates range degradation and rate of climb reduction due to modified aerodynamics.

The final block of code provides an example of computing these penalties using sample aerodynamic data and ice shape factors.

Chapter 64

Algorithms for Supersonic and Transonic Flow Simulation

Below is a Python code snippet that demonstrates solving transonic flow phenomena using high-resolution numerical schemes. The code includes key functions such as the implementation of a high-resolution shock-capturing scheme, the application of boundary conditions, and the main solver loop for a 1D Euler equation setup.

```python
import numpy as np
import matplotlib.pyplot as plt

def initial_conditions(nx):
    """
    Set the initial conditions for the 1D domain.
    :param nx: Number of spatial grid points.
    :return: Initial conditions for density, velocity, and pressure.
    """
    rho = np.ones(nx)
    u = np.zeros(nx)
    p = np.ones(nx) * 101325  # Standard atmospheric pressure

    # Shock front at the center
    rho[int(0.5 * nx):] = 0.125
    u[int(0.5 * nx):] = 0.0
```

```python
    p[int(0.5 * nx):] = 0.1 * 101325

    return rho, u, p

def compute_flux(rho, u, p, gamma=1.4):
    """
    Compute fluxes for the Euler equations.
    :param rho: Density array.
    :param u: Velocity array.
    :param p: Pressure array.
    :param gamma: Specific heat ratio.
    :return: Array of fluxes.
    """
    E = p / (gamma - 1) + 0.5 * rho * u**2  # Total Energy
    F_rho = rho * u
    F_u = rho * u**2 + p
    F_E = u * (E + p)
    return np.array([F_rho, F_u, F_E])

def apply_boundary_conditions(rho, u, p):
    """
    Apply reflective boundary conditions at the domain endpoints.
    :param rho: Density array.
    :param u: Velocity array.
    :param p: Pressure array.
    :return: Updated boundary conditions.
    """
    rho[0] = rho[1]
    u[0] = u[1]
    p[0] = p[1]
    rho[-1] = rho[-2]
    u[-1] = u[-2]
    p[-1] = p[-2]
    return rho, u, p

def euler_solver(nx=100, nt=100, dt=0.0001, dx=0.01):
    """
    Main solver loop for the 1D Euler equations.
    :param nx: Number of spatial grid points.
    :param nt: Number of time steps.
    :param dt: Time step size.
    :param dx: Space step size.
    """
    rho, u, p = initial_conditions(nx)
    gamma = 1.4

    for _ in range(nt):
        rho, u, p = apply_boundary_conditions(rho, u, p)

        # Compute fluxes
        F = compute_flux(rho, u, p, gamma)

        # Update variables using a simple explicit scheme
```

196

```
rho[1:-1] -= dt / dx * (F[0, 1:] - F[0, :-1])
u[1:-1] -= dt / dx * ((F[1, 1:] - F[1, :-1]) / rho[1:-1])
p[1:-1] -= dt / dx * ((F[2, 1:] - F[2, :-1]) - 0.5 *
↪     rho[1:-1] * (u[1:-1]**2 + (p[1:] - p[:-1]) / rho[1:-1]))

    return rho, u, p

# Run the simulation
nx, nt = 200, 500
rho, u, p = euler_solver(nx, nt)

# Plot the results
plt.figure(figsize=(12, 6))
plt.plot(np.linspace(0, 1, nx), rho, label='Density')
plt.plot(np.linspace(0, 1, nx), u, label='Velocity')
plt.plot(np.linspace(0, 1, nx), p / 101325, label='Pressure')
plt.legend()
plt.title('1D Shock Tube Problem')
plt.xlabel('Normalized Position')
plt.show()
```

This code defines and solves the 1D Euler equations for gas dynamics with a shock tube initial condition:

- `initial_conditions` function sets the physical states across the domain before simulation.

- `compute_flux` calculates fluxes for mass, momentum, and energy.

- `apply_boundary_conditions` enforces reflective boundary conditions.

- `euler_solver` iteratively applies the time-stepping scheme to evolve the solution in time.

The script plots the density, velocity, and scaled pressure distribution at the end of the simulation.

197

Chapter 65

Structural Topology Optimization Algorithms

Below is a Python code snippet that exemplifies algorithms for structural topology optimization, focusing on optimal material layout within a predefined design space to achieve superior structural performance. This includes setting up and solving a topology optimization problem using a simple finite element method (FEM) framework.

```python
import numpy as np
import scipy.sparse as sp
import scipy.sparse.linalg as spla

def setup_fem(nelx, nely):
    '''
    Setup global stiffness matrix and other parameters for FEM.
    :param nelx: Number of elements along x.
    :param nely: Number of elements along y.
    :return: Stiffness matrix, Load vector and Boundary conditions.
    '''
    # Initialize stiffness matrix and load vector
    K = sp.lil_matrix((2*(nelx+1)*(nely+1), 2*(nelx+1)*(nely+1)))
    F = np.zeros((2*(nelx+1)*(nely+1), 1))
    U = np.zeros((2*(nelx+1)*(nely+1), 1))

    # Define the element stiffness matrix (4-node square element)
    E = 1.0  # Young's modulus
    nu = 0.3  # Poisson's ratio
```

```python
    k = E / (1 - nu**2) * np.array([[ 1, nu,  0 ],
                                    [ nu,  1,  0 ],
                                    [  0,  0, (1-nu)/2 ]])

    # Assemble the global stiffness matrix
    for elx in range(nelx):
        for ely in range(nely):
            # Element degrees of freedom (DOFs)
            n1 = (nely+1)*elx + ely
            n2 = (nely+1)*(elx+1) + ely
            edof = np.array([2*n1, 2*n1+1, 2*n2, 2*n2+1,
                             2*(n2+1), 2*(n2+1)+1, 2*(n1+1),
                             ↪  2*(n1+1)+1])

            # Add contribution to global stiffness matrix
            for i in range(8):
                for j in range(8):
                    K[edof[i], edof[j]] += k[i%4, j%4]

    # Define load and boundary conditions
    # Load
    F[1, 0] = -1

    # Support (fixed nodes)
    for i in range(nely+1):
        K[2*i, 2*i] = K[2*i+1, 2*i+1] = 1e30  # Large stiffness to
        ↪  simulate fixed boundary

    return K, F, U

def optimize_topology(nelx, nely, vol_frac, penal, rmin,
↪  filter_type='sensitivity'):
    '''
    Perform topology optimization using SIMP (Solid Isotropic
    ↪  Material
    with Penalization) method.
    :param nelx: Number of elements along x.
    :param nely: Number of elements along y.
    :param vol_frac: Volume fraction constraint.
    :param penal: Penalization factor.
    :param rmin: Filter radius.
    :param filter_type: Type of sensitivity filtering ('sensitivity'
    ↪  or 'density').
    :return: Optimized density distribution.
    '''
    K, F, U = setup_fem(nelx, nely)
    x = vol_frac * np.ones(nelx*nely, dtype=float)
    x_old = x.copy()
    dc = np.zeros(nelx*nely, dtype=float)
    while True:
        # FE-Analysis
        U = spla.spsolve(K, F)
```

199

```python
        # Objective and Sensitivity Analysis
        c = 0
        for elx in range(nelx):
            for ely in range(nely):
                # Element DOFs
                n1 = (nely+1)*elx + ely
                n2 = (nely+1)*(elx+1) + ely
                edof = np.array([2*n1, 2*n1+1, 2*n2, 2*n2+1,
                                 2*(n2+1), 2*(n2+1)+1, 2*(n1+1),
                              ↪  2*(n1+1)+1])
                # Element stiffness
                ke = np.array([[ 1, -1, -1,  1],
                               [-1,  1,  1, -1],
                               [-1,  1,  1, -1],
                               [ 1, -1, -1,  1]]) * x[elx*nely + ely]
                              ↪  ** penal
                # Add to global sensivity
                c += 0.5 * np.dot(U[edof].T, ke.dot(U[edof]))
                dc[elx*nely + ely] =
                              ↪  -0.5*U[edof].dot(ke.dot(U[edof]))

        # Sensitivity Filtering
        if filter_type == 'sensitivity':
            dc[:] = sensitivity_filter(dc, nelx, nely, rmin)

        # Optimality Criteria Update
        x = optimality_criteria_update(x, dc, vol_frac)

        # Break if change in design is negligible
        change = np.max(np.abs(x - x_old))
        if change < 0.01:
            break
        x_old = x.copy()

    return x

def sensitivity_filter(dc, nelx, nely, rmin):
    '''
    Apply sensitivity filtering.
    :param dc: Sensitivity values.
    :param nelx: Number of elements along x.
    :param nely: Number of elements along y.
    :param rmin: Filter radius.
    :return: Filtered sensitivity values.
    '''
    # Smoothening based on neighboring elements within rmin
    dc_filtered = np.zeros_like(dc)
    for i in range(nelx):
        for j in range(nely):
            nearby_indices = []
            for k in range(max(i-int(rmin), 0), min(i+int(rmin)+1,
              ↪  nelx)):
```

```python
        for l in range(max(j-int(rmin), 0),
        ↪   min(j+int(rmin)+1, nely)):
            nearby_indices.append(k * nely + l)
        dc_filtered[i*nely + j] = np.mean(dc[nearby_indices])
    return dc_filtered


def optimality_criteria_update(x, dc, vol_frac):
    '''
    Update densities using Optimality Criteria method.
    :param x: Current densities.
    :param dc: Sensitivity values.
    :param vol_frac: Volume constraint.
    :return: Updated densities.
    '''
    # Compute Lagrange multiplier
    l1, l2 = 0, 1e9
    xnew = np.zeros_like(x)
    move = 0.2
    while (l2 - l1)/(l1 + l2) > 1e-3:
        lmid = 0.5 * (l1 + l2)
        xnew[:] = np.maximum(0.0, np.maximum(x-move, np.minimum(1.0,
        ↪   np.minimum(x+move, x * np.sqrt(-dc/lmid)))))
        if sum(xnew) - vol_frac*len(x) > 0:
            l1 = lmid
        else:
            l2 = lmid
    return xnew


# Example case
nelx, nely = 60, 30
vol_frac = 0.4
penal = 3.0
rmin = 1.5

optimized_density = optimize_topology(nelx, nely, vol_frac, penal,
↪   rmin, filter_type='sensitivity')
print("Optimized Density Distribution:", optimized_density)
```

This code covers the key computational elements necessary for topology optimization in structural design:

- `setup_fem` function sets up the finite element model for analyzing structural performance.

- `optimize_topology` utilizes the SIMP (Solid Isotropic Material with Penalization) method to optimize material distribution.

- `sensitivity_filter` applies a filter to ensure smooth sensitivity distribution across the design.

- `optimality_criteria_update` ensures the design adheres to volume constraints while maximizing performance.

The code executes an example of optimizing a 60x30 grid, calibrated to maintain a specified volume fraction while maximizing structural integrity.

Chapter 66

Computational Methods for Turbomachinery Analysis

Below is a Python code snippet that encompasses the core computational elements for simulating the complex flows within turbomachinery components. The code focuses on setting up a simplified model of a compressor or turbine using fundamental fluid dynamics approximations.

```python
import numpy as np
from scipy.optimize import fsolve
import matplotlib.pyplot as plt

def mass_flow_rate(P_in, P_out, T_in, gamma=1.4, R=287):
    '''
    Calculate the mass flow rate through a compressor or turbine.
    :param P_in: Inlet pressure in Pascals.
    :param P_out: Outlet pressure in Pascals.
    :param T_in: Inlet temperature in Kelvin.
    :param gamma: Heat capacity ratio.
    :param R: Specific gas constant for air.
    :return: Mass flow rate in kg/s.
    '''
    rho_in = P_in / (R * T_in)
    a_in = np.sqrt(gamma * R * T_in)
    return rho_in * a_in
```

```python
def isentropic_efficiency(P_in, P_out, T_in, gamma=1.4):
    '''
    Calculate the isentropic efficiency of process.
    :param P_in: Inlet pressure.
    :param P_out: Outlet pressure.
    :param T_in: Inlet temperature.
    :param gamma: Heat capacity ratio.
    :return: Isentropic efficiency (dimensionless).
    '''
    T_out_isentropic = T_in * (P_out / P_in) ** ((gamma - 1) /
    ↪   gamma)
    efficiency = (T_out_isentropic - T_in) / (T_out_isentropic -
    ↪   T_in)
    return efficiency

def solve_pressure_ratio(mass_flow_target, T_in, efficiency,
↪  gamma=1.4, R=287):
    '''
    Solve for the pressure ratio given the mass flow rate target.
    :param mass_flow_target: Desired mass flow rate in kg/s.
    :param T_in: Inlet temperature in Kelvin.
    :param efficiency: System efficiency.
    :param gamma: Heat capacity ratio.
    :param R: Specific gas constant.
    :return: Pressure ratio.
    '''
    def equation(P_ratio):
        P_in = 101325   # Atmospheric pressure in Pascals as
        ↪   reference point
        P_out = P_in * P_ratio
        mass_flow = mass_flow_rate(P_in, P_out, T_in, gamma, R)
        return mass_flow - mass_flow_target

    pressure_ratio = fsolve(equation, 2.0)   # Initial guess of 2.0
    ↪   for solver
    return pressure_ratio[0]

# Example inputs
P_in = 101325   # Inlet pressure in Pascals
Mass_flow_target = 5   # Target mass flow rate in kg/s
T_in = 300   # Inlet temperature in Kelvin
gamma = 1.4   # Heat capacity ratio for air
R = 287   # Specific gas constant for air
efficiency = 0.85   # Assumed efficiency

# Compute pressure ratio
pressure_ratio = solve_pressure_ratio(Mass_flow_target, T_in,
↪   efficiency, gamma, R)

# Output computation results
print(f"Calculated Pressure Ratio: {pressure_ratio:.2f}")
```

```
# Isentropic efficiency and mass flow rate plot
P_out = P_in * pressure_ratio
mass_flow = mass_flow_rate(P_in, P_out, T_in, gamma, R)
efficiency_curve = isentropic_efficiency(P_in, P_out, T_in, gamma)

# Visualization
pressure_ratios = np.linspace(1, 10, 100)
mass_flows = [mass_flow_rate(P_in, P_in * pr, T_in) for pr in
↪    pressure_ratios]
efficiencies = [isentropic_efficiency(P_in, P_in * pr, T_in) for pr
↪    in pressure_ratios]

plt.figure(figsize=(12, 6))
plt.subplot(1, 2, 1)
plt.plot(pressure_ratios, mass_flows, label="Mass Flow Rate")
plt.xlabel("Pressure Ratio")
plt.ylabel("Mass Flow Rate (kg/s)")
plt.title("Mass Flow Rate vs Pressure Ratio")
plt.grid(True)
plt.legend()

plt.subplot(1, 2, 2)
plt.plot(pressure_ratios, efficiencies, label="Isentropic
↪    Efficiency", color='r')
plt.xlabel("Pressure Ratio")
plt.ylabel("Efficiency")
plt.title("Isentropic Efficiency vs Pressure Ratio")
plt.grid(True)
plt.legend()

plt.tight_layout()
plt.show()
```

This code defines several key functions necessary for simulating flow within turbomachinery components:

- mass_flow_rate computes the mass flow rate through turbomachinery based on given inlet conditions and gas properties.

- isentropic_efficiency calculates the isentropic efficiency of a process which is a key performance metric in compressors and turbines.

- solve_pressure_ratio uses the SciPy library's fsolve to find the pressure ratio for a desired mass flow rate, illustrating a typical optimization problem in turbomachinery design.

The final block of code provides examples of how to compute and visualize the impact of pressure ratio on mass flow and efficiency using the outlined functions.

Chapter 67

System Identification Algorithms for Aerospace Vehicles

Below is a Python code snippet that demonstrates the system identification process of developing mathematical models of aerospace systems using experimental or simulated data. The code highlights data preprocessing, implementation of identification algorithms, and validation phases.

```python
import numpy as np
from scipy.optimize import curve_fit
import matplotlib.pyplot as plt

# Sample data generation
def system_response(t, a, b, c):
    '''
    Synthetic system response function mimicking an aerospace
    ↪ system.
    :param t: Time variable.
    :param a, b, c: System parameters.
    :return: System output.
    '''
    return a * np.exp(-b * t) * np.sin(c * t)

# Generate synthetic data
np.random.seed(42)
t_data = np.linspace(0, 10, num=100)
a_true, b_true, c_true = 1.0, 0.2, 2.5
y_data = system_response(t_data, a_true, b_true, c_true) + 0.1 *
    ↪ np.random.normal(size=t_data.size)
```

```python
# System Identification: Curve Fitting
def identify_system(t, y):
    '''
    Identify system parameters using curve fitting.
    :param t: Time data.
    :param y: Observed system data.
    :return: Estimated parameters.
    '''
    def model_func(t, a, b, c):
        return system_response(t, a, b, c)

    popt, _ = curve_fit(model_func, t, y, p0=(1, 0.1, 2))
    return popt

identified_params = identify_system(t_data, y_data)

# Validation: Compare fitted model against actual data
y_fitted = system_response(t_data, *identified_params)

# Plot for visualization
plt.figure(figsize=(10, 6))
plt.plot(t_data, y_data, 'o', label='Observed Data')
plt.plot(t_data, y_fitted, '-', label=f'Fitted Model:
    a={identified_params[0]:.2f}, b={identified_params[1]:.2f},
    c={identified_params[2]:.2f}')
plt.xlabel('Time')
plt.ylabel('System Response')
plt.legend()
plt.title('System Identification through Curve Fitting')
plt.show()

print(f"Identified Parameters: a={identified_params[0]:.2f},
    b={identified_params[1]:.2f}, c={identified_params[2]:.2f}")
```

This code encompasses the fundamental steps in the system identification process:

- `system_response` function generates synthetic data to simulate the behavior of an aerospace system using predefined parameters.

- `identify_system` uses curve fitting techniques to estimate and tune model parameters based on observed data, helping in constructing a mathematical model.

- Validation is performed through comparison between the fitted model's response and the actual observed data, visualized using Matplotlib.

The entire process from data generation to validation is exemplified using a synthetic dataset, reflecting the typical workflow involved in aerospace system modeling using experimental data.

Chapter 68

Computational Modeling of Composite Materials

Below is a Python code snippet that provides a basic implementation of an algorithm to predict the behavior of composite materials used in aerospace structures. The snippet uses a finite element method (FEM) approach to evaluate stress distribution in a composite material characterized by different material properties.

```python
import numpy as np

# Define material properties for composite layers
class CompositeMaterial:
    def __init__(self, modulus, poisson_ratio, thickness):
        self.modulus = modulus
        self.poisson_ratio = poisson_ratio
        self.thickness = thickness

# Set up a simple finite element analysis
def solve_fem(composites, force, area):
    """
    Solves the stress distribution in a composite material using
    ↪    FEM.
    :param composites: List of CompositeMaterial objects.
    :param force: Applied force (N).
    :param area: Cross-sectional area (m^2).
    :return: Stress distribution across layers.
    """

    # Calculate total thickness
```

```python
    total_thickness = sum([c.thickness for c in composites])

    # Calculate stress in each layer
    stress_distribution = []
    for c in composites:
        # Calculate axial stress using simple mechanics
        stress = (force / area) * (c.modulus / (total_thickness *
        ↪   sum(
            [comp.modulus for comp in composites]))) * c.thickness
        stress_distribution.append(stress)

    return stress_distribution

# Example composite material layers
layer1 = CompositeMaterial(70000, 0.3, 0.002)   # E=70GPa, t=2mm
layer2 = CompositeMaterial(150000, 0.25, 0.001)   # E=150GPa, t=1mm

# List of layers forming the composite
composite_layers = [layer1, layer2]

# Define applied force and area
applied_force = 1000   # Newtons
cross_sectional_area = 0.005   # m^2

# Solve for stress distribution
stress_distribution = solve_fem(composite_layers, applied_force,
↪   cross_sectional_area)

# Output results
for idx, stress in enumerate(stress_distribution):
    print(f"Layer {idx + 1} Stress: {stress:.2f} Pa")
```

This code defines the essential components to perform a finite element analysis on composite materials:

- `CompositeMaterial` class captures the modulus of elasticity, Poisson's ratio, and thickness of each composite layer.

- `solve_fem` function calculates and returns the stress distribution across the layers of the composite material by considering applied force and cross-sectional area.

- It demonstrates the implementation using a small set of composite layers with defined material properties, applied force, and cross-sectional area.

The final section of the code provides outputs illustrating the stress distribution across each layer in the composite material through a simple example.

Chapter 69

Unsteady Aerodynamics Simulation Algorithms

Below is a Python code snippet that encompasses the core computational elements for simulating time-dependent aerodynamic phenomena, such as gust response and dynamic stall.

```python
import numpy as np

def simulate_gust_response(aircraft_state, gust_intensity,
↪    time_step, duration):
    '''
    Simulate the aircraft's response to a gust of wind.
    :param aircraft_state: Initial state vector [speed,
    ↪    angle_of_attack, etc.]
    :param gust_intensity: Intensity of the gust impacting the
    ↪    aircraft.
    :param time_step: Time increment for simulation.
    :param duration: Total time duration for the simulation.
    :return: Time series of state vectors.
    '''
    time_points = int(duration / time_step)
    state_history = np.zeros((time_points, len(aircraft_state)))
    state_history[0] = aircraft_state

    for t in range(1, time_points):
        # Simplified dynamics update due to gust
        speed, angle_of_attack = state_history[t-1]
        speed += gust_intensity * np.cos(angle_of_attack) *
        ↪    time_step
```

```python
            angle_of_attack += gust_intensity * np.sin(angle_of_attack)
            ↪   / speed * time_step

            # Store calculated state
            state_history[t] = [speed, angle_of_attack]

    return state_history

def simulate_dynamic_stall(airfoil_state, alpha_dynamic,
↪ alpha_stall, step, max_iterations):
    '''
    Simulate the dynamic stall effects on an airfoil.
    :param airfoil_state: Initial aerodynamic state variables.
    :param alpha_dynamic: Dynamic angle of attack.
    :param alpha_stall: Critical angle of attack for stall.
    :param step: Time step for integration.
    :param max_iterations: Maximum iterations to simulate.
    :return: List of aerodynamic coefficients over time.
    '''
    lift_coefficient = airfoil_state['CL0']
    drag_coefficient = airfoil_state['CD0']

    cl_time_series = [lift_coefficient]
    cd_time_series = [drag_coefficient]

    for i in range(max_iterations):
        if alpha_dynamic > alpha_stall:
            # Induce stall effects
            df_lift = -0.1 * (alpha_dynamic - alpha_stall) * step
            df_drag = 0.1 * (alpha_dynamic - alpha_stall) * step
            frequency_response = 0.1  # Effect damping post-stall
        else:
            df_lift = 0.05 * (alpha_stall - alpha_dynamic) * step
            df_drag = -0.05 * (alpha_stall - alpha_dynamic) * step
            frequency_response = 0.05

        # Update coefficients
        lift_coefficient += df_lift * frequency_response
        drag_coefficient += df_drag * frequency_response

        cl_time_series.append(lift_coefficient)
        cd_time_series.append(drag_coefficient)

    return cl_time_series, cd_time_series

# Example variables for simulation
aircraft_initial_state = [250.0, 0.05]  # Example: speed in m/s,
↪ angle of attack in radians
gust_intensity = 5.0
time_step = 0.1
simulation_duration = 10.0

airfoil_initial_conditions = {'CL0': 0.5, 'CD0': 0.02}
```

```
alpha_dynamic = 0.15  # rad
alpha_stall = 0.12  # rad
simulation_steps = 200

# Perform simulations
gust_response = simulate_gust_response(aircraft_initial_state,
↪  gust_intensity, time_step, simulation_duration)
dynamic_stall_response =
↪  simulate_dynamic_stall(airfoil_initial_conditions,
↪  alpha_dynamic, alpha_stall, time_step, simulation_steps)

# Demo outputs
print("Gust Response:", gust_response[-1])
print("Dynamic Stall Response (final CL, CD):",
↪  dynamic_stall_response[0][-1], dynamic_stall_response[1][-1])
```

This code defines several key functions necessary for simulating time-dependent aerodynamic phenomena:

- `simulate_gust_response` function models how an aircraft responds to wind gusts, updating its state over time.

- `simulate_dynamic_stall` implements a basic model for dynamic stall, showing changes in lift and drag coefficients.

- Both simulations return time-series data that illustrates system responses to aerodynamic changes.

The final block of code provides examples and demonstration outputs for gust response and dynamic stall simulations using predefined conditions.

Chapter 70

Algorithms for Mars Entry, Descent, and Landing Simulations

Below is a Python code snippet that encompasses the core computational elements for simulating Entry, Descent, and Landing (EDL) sequences on Mars, taking into account atmospheric and terrain interactions.

```python
import numpy as np
from scipy.integrate import solve_ivp

def mars_atmospheric_density(altitude):
    '''
    Calculate the atmospheric density at a given altitude on Mars.
    :param altitude: Altitude above Mars' surface in meters.
    :return: Atmospheric density in kg/m^3.
    '''
    # Constants for Mars atmosphere model
    base_density = 0.020   # kg/m^3 at surface
    scale_height = 11100   # meters

    return base_density * np.exp(-altitude / scale_height)

def edl_dynamics(t, y, area, mass, Cd, gravity):
    '''
    Governs the dynamics of the spacecraft during EDL.
    :param t: Time variable (needed for solve_ivp).
    :param y: State variables [altitude, velocity].
    :param area: Cross-sectional area of the spacecraft.
    :param mass: Mass of the spacecraft.
```

```
    :param Cd: Drag coefficient.
    :param gravity: Martian gravitational acceleration.
    :return: Derivative [d(altitude)/dt, d(velocity)/dt].
    '''

    altitude, velocity = y

    if altitude <= 0:
        return [0, 0]    # No further motion once on ground

    density = mars_atmospheric_density(altitude)
    drag_force = 0.5 * density * velocity**2 * area * Cd
    acceleration_due_to_drag = drag_force / mass

    # Equations of motion
    dydt = [velocity, -gravity + acceleration_due_to_drag]

    return dydt

def simulate_edl(initial_altitude, initial_velocity, area, mass, Cd,
↪   gravity):
    '''
    Simulate the Entry, Descent, and Landing sequence.
    :param initial_altitude: Starting altitude for EDL.
    :param initial_velocity: Starting velocity for EDL.
    :param area: Cross-sectional area of the spacecraft.
    :param mass: Mass of the spacecraft.
    :param Cd: Drag coefficient.
    :param gravity: Martian gravitational acceleration.
    :return: Results of the simulation including time, altitude, and
    ↪   velocity.
    '''

    initial_conditions = [initial_altitude, initial_velocity]
    t_span = (0, 1000)   # Simulate for 1000 seconds

    solution = solve_ivp(edl_dynamics, t_span, initial_conditions,
    ↪   args=(area, mass, Cd, gravity), dense_output=True)

    return solution.t, solution.y

# Parameters for the Mars EDL simulation
initial_altitude = 125000   # in meters
initial_velocity = -5500   # in meters per second (descending)
cross_sectional_area = 15   # in square meters
mass = 900   # in kg
drag_coefficient = 1.5   # assumption for descent module
martian_gravity = 3.71   # in meters per second squared

# Executing the simulation
time, results = simulate_edl(initial_altitude, initial_velocity,
↪   cross_sectional_area, mass, drag_coefficient, martian_gravity)

# Extract altitude and velocity results
altitude = results[0]
```

```
velocity = results[1]

for t, alt, vel in zip(time, altitude, velocity):
    print(f"Time: {t:.2f}, Altitude: {alt:.2f}, Velocity:
    ↪ {vel:.2f}")
```

This code defines several key functions necessary for simulating Mars EDL sequences:

- `mars_atmospheric_density` calculates the atmospheric density at a given altitude using an exponential model.

- `edl_dynamics` defines the differential equations for altitude and velocity during EDL, influenced by drag and gravity.

- `simulate_edl` executes the simulation using these dynamics to solve the system over time.

The final block of code sets parameters for a hypothetical EDL mission and prints the simulation results, displaying changes in altitude and velocity over time.

Chapter 71

Computational Methods for Spacecraft Thermal Control

Below is a Python code snippet that involves key computational methods for designing and analyzing thermal control systems in spacecraft, including temperature distribution calculation, active thermal management strategies, and thermal model validation.

```python
import numpy as np

def simulate_temperature_distribution(material_properties,
↪    boundary_conditions, initial_state, time_steps):
    '''
    Simulates temperature distribution in a spacecraft component.
    :param material_properties: Dictionary containing thermal
    ↪    properties such as conductivity and specific heat.
    :param boundary_conditions: Temperature and heat flux conditions
    ↪    at the boundaries.
    :param initial_state: Initial temperature distribution.
    :param time_steps: Number of time steps for simulation.
    :return: Array of temperature distributions over time.
    '''
    # Initialize temperature array
    temperature = initial_state.copy()

    # Time integration loop
    for _ in range(time_steps):
        # Simplified heat conduction update (finite difference)
        temperature[1:-1] = (temperature[:-2] +
                             temperature[2:] +
```

```
                    material_properties['conductivity'] *
                    (temperature[1:-1] -
                    ↪   boundary_conditions['ambient'])) /
                    ↪   2

        # Apply boundary conditions
        temperature[0] = boundary_conditions['left']
        temperature[-1] = boundary_conditions['right']

    return temperature

def active_thermal_control(temperature, set_points, control_action):
    '''
    Adjusts heating or cooling elements to maintain off-nominal
    ↪   temperatures within desired set points.
    :param temperature: Current temperature distribution.
    :param set_points: Desired temperature range.
    :param control_action: Function representing control strategy,
    ↪   such as PID control.
    :return: Adjusted temperatures after control action.
    '''

    adjustments = control_action(temperature, set_points)
    return temperature + adjustments

def validate_thermal_model(simulation_result, experimental_data):
    '''
    Validates the thermal model by comparing simulation results to
    ↪   experimental measurements.
    :param simulation_result: Predicted temperatures from
    ↪   simulation.
    :param experimental_data: Actual temperatures recorded during
    ↪   experiments.
    :return: Validation metrics such as mean absolute error.
    '''

    mae = np.mean(np.abs(simulation_result - experimental_data))
    return mae

# Example material properties and boundary conditions
material_props = {'conductivity': 0.5, 'specific_heat': 900}
boundary_conditions = {'left': 293, 'right': 293, 'ambient': 273}

# Initial distribution
initial_temp = np.array([300] * 10)

# Time steps for transient simulation
time_steps = 1000

# Simulate temperature distribution
simulated_temperatures =
↪   simulate_temperature_distribution(material_props,
↪   boundary_conditions, initial_temp, time_steps)

# Define a simple control strategy
```

218

```
def simple_control_action(current_temps, target):
    return np.clip(target - current_temps, -1, 1)

# Set desired temperature set points
desired_set_points = np.array([290] * 10)

# Apply active thermal control
controlled_temperatures =
↪   active_thermal_control(simulated_temperatures,
↪   desired_set_points, simple_control_action)

# Dummy experimental data for validation
experimental_data = np.array([295] * 10)

# Validate model
validation_error = validate_thermal_model(controlled_temperatures,
↪   experimental_data)

print("Simulated Temperatures:", simulated_temperatures)
print("Controlled Temperatures:", controlled_temperatures)
print("Validation Error (MAE):", validation_error)
```

This code defines several key functions for modeling and controlling spacecraft thermal environments:

- `simulate_temperature_distribution` simulates the temperature profile using a simplified heat conduction calculation over specified time steps.

- `active_thermal_control` modifies the temperature distribution by applying a user-defined control strategy to meet designated set points.

- `validate_thermal_model` compares simulation results with experimental data to assess predictive accuracy.

The example initializes material properties, boundary conditions, and initial temperatures, applies control adjustments to the temperature distribution, and validates the model against dummy experimental data.

Chapter 72

Free-Surface Flow Simulation Algorithms

Below is a Python code snippet that encompasses the core computational elements for simulating flows with free surfaces, which are relevant to applications like fuel slosh in tanks and flotation behavior.

```python
import numpy as np
import matplotlib.pyplot as plt
from scipy.integrate import solve_ivp

def free_surface_flow(t, y, params):
    '''
    Calculate the rate of change of state variables for free-surface
    ↪ flow.
    :param t: Current time.
    :param y: Current state vector.
    :param params: Dictionary containing fluid and tank parameters.
    :return: Rate of change of state vector.
    '''
    g = params['g']  # gravitational acceleration
    A = params['A']  # cross-sectional area
    h = y[0]  # liquid height

    dhdt = -(g / A) * h  # Example: Simplified linear drainage
    return [dhdt]

def simulate_flow(params, t_span, y0):
    '''
    Simulate the free-surface flow in a container.
    :param params: Dictionary containing parameters for the
    ↪ simulation.
```

```
    :param t_span: Tuple indicating the start and end of the
    ↪  simulation time.
    :param y0: Initial state vector.
    :return: Time points and state vector.
    '''

    sol = solve_ivp(free_surface_flow, t_span, y0, args=(params,),
    ↪  dense_output=True)

    if sol.success:
        return sol.t, sol.y.T
    else:
        raise RuntimeError('Simulation failed')

# Parameters
parameters = {
    'g': 9.81,   # m/s^2, gravitational acceleration
    'A': 1.0     # m^2, cross-sectional area
}

# Time range for the simulation
time_span = (0, 10)   # seconds

# Initial conditions: Initial height of the liquid
initial_conditions = [1.0]   # meters

# Run the simulation
time_points, heights = simulate_flow(parameters, time_span,
↪  initial_conditions)

# Plotting the results
plt.figure(figsize=(10, 6))
plt.plot(time_points, heights, label='Liquid Height (m)')
plt.xlabel('Time (s)')
plt.ylabel('Height (m)')
plt.title('Free-Surface Flow Simulation')
plt.legend()
plt.grid()
plt.show()
```

This code defines several key functions necessary for simulating free-surface flow:

- `free_surface_flow` function models the rate of change of liquid height in a container with a free surface, influenced by gravitational forces.

- `simulate_flow` function uses numerical integration to solve the differential equation governing the fluid flow, yielding time-dependent profiles of liquid height.

- Visualization of the simulation results is facilitated using `matplotlib`, providing insights into the dynamic behavior of the free surface.

The Python code provides a simple example of simulating the liquid drainage from a tank under gravity, offering a foundation for more complex models involving fluid slosh or flotation phenomena.

Chapter 73

Algorithms for Flexible Multibody Dynamics

Below is a Python code snippet that represents a basic implementation of algorithms to simulate the interconnected dynamics of rigid and flexible bodies using multibody dynamics principles. The code includes a class definition to encapsulate the dynamics and numerical integration steps for simulating the motion of the bodies.

```python
import numpy as np
from scipy.integrate import odeint

class RigidFlexibleBodyDynamics:
    def __init__(self, m_rigid, k_flexible, c_flexible):
        '''
        Initialize the dynamics of rigid and flexible bodies.
        :param m_rigid: Mass of the rigid body.
        :param k_flexible: Stiffness of the flexible element.
        :param c_flexible: Damping of the flexible element.
        '''
        self.m_rigid = m_rigid
        self.k_flexible = k_flexible
        self.c_flexible = c_flexible

    def equations_of_motion(self, state, t):
        '''
        Define the equations of motion for the system.
        :param state: Current state vector (positions and
        ↪    velocities).
        :param t: Time.
        :return: Derivatives of the state vector.
```

```
    '''
    x, v = state
    a = -(self.k_flexible / self.m_rigid) * x - (self.c_flexible
    ↪ / self.m_rigid) * v
    return [v, a]

def simulate(self, initial_conditions, time_span):
    '''
    Simulate the dynamics over a given time span.
    :param initial_conditions: Initial state [x0, v0].
    :param time_span: Time span for simulation.
    :return: Simulated positions and velocities over time.
    '''
    result = odeint(self.equations_of_motion,
    ↪ initial_conditions, time_span)
    return result

# Example usage of the class to simulate dynamics
mass_of_rigid_body = 10.0   # kg
stiffness_of_flexible_body = 50.0   # N/m
damping_of_flexible_body = 5.0   # Ns/m

initial_conditions = [0.1, 0.0]   # Initial displacement and velocity
time_span = np.linspace(0, 10, 100)   # Time from 0 to 10 seconds

simulation = RigidFlexibleBodyDynamics(mass_of_rigid_body,
↪ stiffness_of_flexible_body, damping_of_flexible_body)
result = simulation.simulate(initial_conditions, time_span)

# Results
positions = result[:, 0]
velocities = result[:, 1]

print("Positions:\n", positions)
print("Velocities:\n", velocities)
```

This code illustrates:

- The `RigidFlexibleBodyDynamics` class, which models the interaction of a rigid body with a flexible element through stiffness and damping.

- The `equations_of_motion` method defines the system's dynamic equations, combining the effects of stiffness and damping forces on the body.

- The `simulate` function utilizes numerical integration to compute the time evolution of the system's state over a specified time span.

- An example block provides a practical case of simulating the motion of the system using defined parameters including mass, stiffness, and damping.

The simulation outputs the positions and velocities of the rigid body over time, showcasing the dynamics resulting from the interplay between rigid and flexible body components.

Chapter 74

Optimal Sensor Placement Algorithms

Below is a Python code snippet that demonstrates an algorithm for determining optimal sensor placement to maximize information gain, which can be crucial for aerospace applications such as structural health monitoring or environmental data collection.

```python
import numpy as np
from scipy.optimize import minimize

def information_gain(sensor_positions, model, data_covariance):
    '''
    Calculate the negative information gain for a given set of
    ↪ sensor positions.
    :param sensor_positions: Array of sensor positions.
    :param model: Function representing the system model.
    :param data_covariance: Covariance matrix of the data.
    :return: Negative information gain value.
    '''
    # Compute the model predictions at sensor positions
    predictions = model(sensor_positions)
    # Compute the covariance of the predictions
    prediction_covariance = np.cov(predictions, rowvar=False)
    # Calculate the information gain (negative for minimization)
    info_gain = np.trace(prediction_covariance @
    ↪ np.linalg.inv(data_covariance))

    return -info_gain

def example_model(sensor_positions):
    '''
```

```
    Example system model that predicts some values based on sensor
    ↪ positions.
    :param sensor_positions: Array of sensor positions.
    :return: Predicted data array.
    '''
    # For demonstration, let's assume a quadratic form
    return sensor_positions ** 2 + 5 * sensor_positions + 3

# Initial sensor positions (for demonstration)
initial_positions = np.array([0.0, 5.0, 10.0])
# Example data covariance matrix
data_covariance_matrix = np.array([[1.0, 0.2, 0.1],
                                   [0.2, 1.0, 0.3],
                                   [0.1, 0.3, 1.0]])

# Optimization to determine best sensor placement
result = minimize(information_gain, initial_positions,
↪ args=(example_model, data_covariance_matrix),
               method='L-BFGS-B', bounds=[(0, 10), (0, 10), (0,
               ↪ 10)])

print("Optimal Sensor Positions:", result.x)
print("Maximum Information Gain:", -result.fun)
```

In this code snippet, we define the core computational components necessary for optimal sensor placement:

- `information_gain` function calculates the negative information gain, which is the objective for minimization in the optimization process.

- `example_model` is a placeholder system model representing
  how sensor data might be generated based on position. It
  can be replaced with a real model specific to the application.

- The optimization uses the `minimize` function from SciPy, applying the L-BFGS-B method to handle the non-linear and
  bound-constrained nature typical in sensor placement problems.

The output of this code lists the computed optimal sensor positions and the associated maximum information gain, demonstrating the algorithm's application.

Chapter 75

Computational Methods for Turbulence Control

Below is a Python code snippet that demonstrates the core computational elements of a turbulence control system, including airflow simulation, feedback control implementation, and adjustment for aerodynamic performance enhancement.

```python
import numpy as np
import matplotlib.pyplot as plt
from scipy.integrate import solve_ivp

def airflow_simulation(v_inf, angle_of_attack):
    '''
    Simulate the airflow over an airfoil with given velocity and
    ↪  angle of attack.
    :param v_inf: Freestream velocity.
    :param angle_of_attack: Angle of attack in degrees.
    :return: Lift and drag coefficients.
    '''
    # Simplified model; replace with detailed CFD model for accuracy
    cl = 2 * np.pi * np.radians(angle_of_attack)
    cd = cl**2 / (np.pi * 4)   # Aspect ratio = 4 for simplicity

    return cl, cd

def feedback_control(target_cl, current_cl):
    '''
    Calculate the necessary control adjustment to reach the target
    ↪  lift coefficient.
```

```python
    :param target_cl: Target lift coefficient.
    :param current_cl: Current lift coefficient.
    :return: Control adjustment angle.
    '''
    k_p = 0.1   # Proportional gain
    error = target_cl - current_cl
    adjustment = k_p * error

    return adjustment

def control_loop(v_inf, desired_lift, initial_aoa):
    '''
    Run the control loop to achieve desired aerodynamic performance.
    :param v_inf: Freestream velocity.
    :param desired_lift: Desired lift coefficient.
    :param initial_aoa: Initial angle of attack.
    :return: History of angles of attack and lift coefficients.
    '''
    aoa_history = [initial_aoa]
    cl_history = []

    for _ in range(100):   # Run for 100 iterations
        current_aoa = aoa_history[-1]
        current_cl, _ = airflow_simulation(v_inf, current_aoa)
        cl_history.append(current_cl)

        adjustment = feedback_control(desired_lift, current_cl)
        new_aoa = current_aoa + adjustment

        aoa_history.append(new_aoa)
        if abs(desired_lift - current_cl) < 0.001:   # Stop if within
        ↪  tolerance
            break

    return aoa_history, cl_history

# Example usage
v_inf = 100   # Freestream velocity in m/s
desired_lift = 1.2
initial_aoa = 5   # Initial angle of attack in degrees

aoa_history, cl_history = control_loop(v_inf, desired_lift,
↪  initial_aoa)

# Plot results
plt.figure()
plt.plot(cl_history)
plt.xlabel('Iteration')
plt.ylabel('Lift Coefficient (Cl)')
plt.title('Lift Coefficient Convergence')
plt.grid()
plt.show()
```

```
plt.figure()
plt.plot(aoa_history)
plt.xlabel('Iteration')
plt.ylabel('Angle of Attack (degrees)')
plt.title('Angle of Attack Adjustment')
plt.grid()
plt.show()
```

This code defines and executes a basic control loop to achieve a target lift coefficient:

- `airflow_simulation` function simulates the aerodynamic properties over an airfoil at a specific angle of attack.

- `feedback_control` computes the control adjustment necessary to align the current lift with the desired lift.

- `control_loop` runs the iterative process of adjusting angle of attack to achieve the specified aerodynamic performance.

The final plots visualize the convergence of the lift coefficient and the corresponding changes in angle of attack over successive iterations.

Chapter 76

Algorithms for Air Traffic Flow Management

Below is a Python code snippet that encompasses the core computational elements of air traffic flow management, including data structures for flight details, algorithms for conflict detection and resolution, and optimization to minimize delays and maximize safety.

```python
import numpy as np
import pandas as pd
from scipy.optimize import minimize

class Aircraft:
    def __init__(self, flight_id, current_position, velocity, path):
        self.flight_id = flight_id
        self.current_position = np.array(current_position)
        self.velocity = np.array(velocity)
        self.path = np.array(path)

def generate_aircraft_data(num_flights):
    '''
    Generate random aircraft data for simulation.
    :param num_flights: Number of flights.
    :return: List of Aircraft objects.
    '''
    aircrafts = []
    for i in range(num_flights):
        flight_id = f"Flight_{i}"
        current_position = np.random.rand(2) * 100   # Random initial
        ↪   position
```

```
            velocity = (np.random.rand(2) - 0.5) * 10   # Random
            ↳   velocity
            path = np.array([current_position + np.random.rand(2) * 1000
            ↳   for _ in range(5)])
            aircrafts.append(Aircraft(flight_id, current_position,
            ↳   velocity, path))
        return aircrafts

def predict_positions(aircraft, time_horizon):
    '''
    Predict future positions of an aircraft.
    :param aircraft: An Aircraft object.
    :param time_horizon: Time period to predict over.
    :return: Predicted positions array.
    '''
    return [aircraft.current_position + aircraft.velocity * t for t
    ↳   in range(1, time_horizon + 1)]

def detect_conflicts(aircrafts, time_horizon, separation_distance):
    '''
    Detect potential conflicts between aircraft over a time horizon.
    :param aircrafts: List of Aircraft objects.
    :param time_horizon: Time horizon to check for conflicts.
    :param separation_distance: Minimum safe separation distance.
    :return: List of conflicting pairs.
    '''
    conflicts = []
    for i, ac1 in enumerate(aircrafts):
        future_positions_1 = predict_positions(ac1, time_horizon)
        for j, ac2 in enumerate(aircrafts[i+1:], i+1):
            future_positions_2 = predict_positions(ac2,
            ↳   time_horizon)
            for t in range(time_horizon):
                distance = np.linalg.norm(future_positions_1[t] -
                ↳   future_positions_2[t])
                if distance < separation_distance:
                    conflicts.append((ac1.flight_id, ac2.flight_id,
                    ↳   t))
    return conflicts

def resolve_conflicts(aircrafts, conflicts):
    '''
    Resolve conflicts by adjusting velocities.
    :param aircrafts: List of Aircraft objects.
    :param conflicts: List of conflicts to resolve.
    :return: Adjusted velocities for each aircraft to avoid
    ↳   conflicts.
    '''
    def objective_function(v_changes):
        cost = 0
        for i, ac in enumerate(aircrafts):
            ac.velocity = ac.velocity + v_changes[2*i:2*i+2]
            # Z function assesses total deviation from optimal paths
```

```
        for conflict in conflicts:
            ac1 = next(ac for ac in aircrafts if ac.flight_id ==
            ↪ conflict[0])
            ac2 = next(ac for ac in aircrafts if ac.flight_id ==
            ↪ conflict[1])
            pred_pos1 = predict_positions(ac1, conflict[2]+1)[-1]
            pred_pos2 = predict_positions(ac2, conflict[2]+1)[-1]
            dist = np.linalg.norm(pred_pos1 - pred_pos2)
            # Penalty if still within conflicted state
            if dist < separation_distance:
                cost += 1000 * (separation_distance - dist)
            # Cost for total deviation or speed
        return cost

    initial_guess = np.zeros(2 * len(aircrafts))
    result = minimize(objective_function, initial_guess,
    ↪ method='SLSQP')
    for i, ac in enumerate(aircrafts):
        ac.velocity += result.x[2*i:2*i+2]

# Parameters and Simulation
num_flights = 10
time_horizon = 5
separation_distance = 5.0

aircrafts = generate_aircraft_data(num_flights)
conflicts = detect_conflicts(aircrafts, time_horizon,
↪ separation_distance)

if conflicts:
    print("Conflicts detected:", conflicts)
    resolve_conflicts(aircrafts, conflicts)
    print("Conflicts resolved:")
else:
    print("No conflicts detected.")
```

This code defines several key functions necessary for air traffic flow management:

- **Aircraft** class models individual aircraft with fields for flight ID, current position, velocity, and planned path.

- **generate_aircraft_data** creates a list of aircraft with randomized initial conditions for simulation purposes.

- **predict_positions** uses kinematics to estimate future positions based on current velocity and time horizon.

- **detect_conflicts** identifies proximity breaches within a specified time horizon, outputting conflicting flight pairs.

- `resolve_conflicts` modifies velocities to maintain safe separation using optimization techniques minimizing overall deviation.

The final block runs the simulation and demonstrates detecting and resolving conflicts among sample flights.

Chapter 77

Finite Element Analysis of Fracture Mechanics

Below is a Python code snippet that encompasses the core computational elements for predicting crack initiation and growth in aerospace materials using fracture mechanics principles and a simple damage accumulation model.

```python
import numpy as np

def stress_intensity_factor(K_ic, sigma, a):
    '''
    Calculate the stress intensity factor (K) for a crack.
    :param K_ic: Fracture toughness of the material.
    :param sigma: Applied stress.
    :param a: Crack length.
    :return: Stress intensity factor.
    '''
    # Assume a simple calculation for a central crack in an infinite
    ↪    plate
    return sigma * np.sqrt(np.pi * a)

def crack_growth_rate(c, m, delta_K):
    '''
    Calculate the crack growth rate using Paris' law.
    :param c: Paris' law constant.
    :param m: Paris' law exponent.
    :param delta_K: Stress intensity factor range.
    :return: Crack growth rate.
    '''
```

```python
    return c * (delta_K ** m)

def simulate_crack_growth(K_ic, sigma, c, m, a_initial, a_final,
↪ da):
    '''
    Simulate crack growth from an initial to a final length.
    :param K_ic: Fracture toughness.
    :param sigma: Applied stress.
    :param c: Paris' law constant.
    :param m: Paris' law exponent.
    :param a_initial: Initial crack length.
    :param a_final: Final crack length.
    :param da: Increment in crack length for each step.
    :return: Array of crack lengths over time.
    '''
    a = a_initial
    crack_lengths = [a]

    while a < a_final:
        delta_K = stress_intensity_factor(K_ic, sigma, a)
        if delta_K >= K_ic:
            print(f"Crack unstable at length {a}")
            break

        da_dt = crack_growth_rate(c, m, delta_K)
        a += da_dt * da
        crack_lengths.append(a)

    return np.array(crack_lengths)

# Example parameters for simulation
K_ic_example = 50.0  # MPam
sigma_example = 200.0  # MPa
c_example = 1e-12
m_example = 3.0
a_initial_example = 0.01  # meters
a_final_example = 0.1  # meters
da_example = 0.001  # meters per cycle

# Run crack growth simulation
crack_sizes = simulate_crack_growth(K_ic_example, sigma_example,
↪ c_example, m_example, a_initial_example, a_final_example,
↪ da_example)

print("Crack lengths over time:", crack_sizes)
```

This code defines several key functions necessary for simulating crack growth in aerospace materials:

- **stress_intensity_factor** function calculates the stress intensity factor for a given crack length and applied stress.

236

- `crack_growth_rate` evaluates the rate of crack growth using Paris' law, a standard approach in fracture mechanics.

- `simulate_crack_growth` iteratively predicts the crack length over time by applying the stress intensity and crack growth rate calculations.

The final block of code provides an example of crack growth simulation over specified conditions and prints the crack lengths at various steps.

Chapter 78

Computational Methods for Bioaerospace Engineering

Below is a Python code snippet that demonstrates a basic framework for modeling bio-inspired structures in aerospace applications. It uses bio-inspired algorithms to optimize structural designs, mimicking mechanical efficiency found in natural biological systems.

```python
import numpy as np
from scipy.optimize import minimize
import matplotlib.pyplot as plt

def bioinspired_design(x):
    '''
    Objective function for bio-inspired structural design
    ↪    optimization.
    :param x: Design variables (e.g., thicknesses or material
    ↪    properties).
    :return: Efficiency metric (to be minimized).
    '''
    # Example: weighted sum of the structure's mass and compliance
    mass = np.sum(x**2)
    compliance = np.sum((1/x)**2)
    efficiency = mass + compliance

    return efficiency
```

```
def constraint_eq(x):
    '''
    Example equality constraint for the design problem.
    :param x: Design variables.
    :return: Constraint value (should be zero for a valid solution).
    '''
    return np.sum(x) - 10   # Example constraint: sum of all design
    ↪  variables equals 10

def constraint_ineq(x):
    '''
    Example inequality constraint for the design problem.
    :param x: Design variables.
    :return: Constraint value (should be non-negative for a valid
    ↪  solution).
    '''
    return x - 0.1   # Example constraint: all design variables
    ↪  should be greater than 0.1

# Initial design
x0 = np.ones(5)   # Example: start with 5 design variables all
↪  initialized to 1

# Run the optimization
constraints = [{'type': 'eq', 'fun': constraint_eq},
               {'type': 'ineq', 'fun': constraint_ineq}]
result = minimize(bioinspired_design, x0, constraints=constraints,
↪  method='SLSQP')

# Display results
print("Optimal Design:", result.x)
print("Objective Function Value:", bioinspired_design(result.x))

# Plotting the optimal design
plt.figure()
plt.bar(range(len(result.x)), result.x)
plt.xlabel('Design Variable Index')
plt.ylabel('Optimized Value')
plt.title('Optimized Bio-Inspired Design Variables')
plt.show()
```

This code outlines a set of basic functions and an optimization process to develop a bio-inspired design:

- **bioinspired_design** serves as the objective function which combines metrics such as structural mass and mechanical compliance for optimization.

- **constraint_eq** is an example of an equality constraint ensuring that the sum of design variables meets a specific requirement.

- `constraint_ineq` provides an inequality constraint ensuring all design variables maintain a minimum value.

The optimization is performed using `scipy.optimize.minimize` with Sequential Least Squares Programming (SLSQP) to meet both equality and inequality constraints, showcasing how nature-inspired efficiencies can guide structural design in aerospace engineering.

Chapter 79

Hybrid RANS/LES Turbulence Modeling Algorithms

Below is a Python code snippet that encompasses the core computational elements of hybrid RANS/LES turbulence modeling, including the setup of the computational domain, implementation of the hybrid turbulence model, and the solver execution for simulation.

```python
import numpy as np

def create_domain(grid_points, length, width):
    '''
    Create the computational domain grid.
    :param grid_points: Number of grid points.
    :param length: Length of the domain.
    :param width: Width of the domain.
    :return: Grid mesh points.
    '''
    x = np.linspace(0, length, grid_points)
    y = np.linspace(0, width, grid_points)
    mesh = np.meshgrid(x, y)
    return mesh

def hybrid_turbulence_model(u, v, rans_coef=0.5, les_coef=0.5):
    '''
    Apply a hybrid RANS/LES model for turbulence simulation.
    :param u: Velocity field in x-direction.
    :param v: Velocity field in y-direction.
```

```python
    :param rans_coef: Coefficient for RANS contribution.
    :param les_coef: Coefficient for LES contribution.
    :return: Hybrid turbulence field.
    '''
    rans_turbulence = rans_coef * calculate_rans_turbulence(u, v)
    les_turbulence = les_coef * calculate_les_turbulence(u, v)
    return rans_turbulence + les_turbulence

def calculate_rans_turbulence(u, v):
    '''
    Calculate RANS turbulence effects.
    :param u: Velocity field in x-direction.
    :param v: Velocity field in y-direction.
    :return: RANS turbulence field.
    '''
    # Placeholder computation
    return np.mean(u) + np.mean(v)

def calculate_les_turbulence(u, v):
    '''
    Calculate LES turbulence effects.
    :param u: Velocity field in x-direction.
    :param v: Velocity field in y-direction.
    :return: LES turbulence field.
    '''
    # Placeholder computation
    return np.std(u) + np.std(v)

def solve_flow(mesh, initial_conditions, time_steps, dt):
    '''
    Execute the solver for the hybrid turbulence simulation.
    :param mesh: Computational domain.
    :param initial_conditions: Initial velocity conditions.
    :param time_steps: Number of time steps to simulate.
    :param dt: Time-step size.
    :return: Final velocity field.
    '''
    u_field, v_field = initial_conditions
    for _ in range(time_steps):
        hybrid_turbulence = hybrid_turbulence_model(u_field,
        ↪ v_field)
        u_field += dt * hybrid_turbulence
        v_field += dt * hybrid_turbulence
    return u_field, v_field

# Setup simulation parameters
grid_points = 100
length = 10.0
width = 5.0
initial_conditions = (np.zeros((grid_points, grid_points)),
↪ np.zeros((grid_points, grid_points)))
time_steps = 100
dt = 0.01
```

```
# Create computational domain and solve
domain_mesh = create_domain(grid_points, length, width)
final_u, final_v = solve_flow(domain_mesh, initial_conditions,
↪   time_steps, dt)

print("Final u-field:", final_u)
print("Final v-field:", final_v)
```

This code defines several key functions necessary for the implementation of a hybrid RANS/LES turbulence model:

- `create_domain` function establishes the computational grid for the simulation domain.

- `hybrid_turbulence_model` applies a hybrid approach, combining RANS and LES turbulence effects.

- `calculate_rans_turbulence` computes the turbulence effects using the RANS model.

- `calculate_les_turbulence` assesses the turbulence contributions from the LES model.

- `solve_flow` iteratively solves the flow field over specified time steps incorporating hybrid turbulence effects.

The implementation provides a simplified framework for modeling turbulence phenomena using hybrid RANS/LES techniques, suitable for expanding into more detailed simulations.

Chapter 80

Algorithms for Spacecraft GNC under Uncertainties

Below is a Python code snippet that encompasses the core computational elements of robust guidance, navigation, and control (GNC) algorithms that handle model uncertainties and external disturbances. This includes state estimation, control input calculation, and disturbance rejection using Kalman filters and PID control strategies.

```python
import numpy as np

class KalmanFilter:
    def __init__(self, A, B, H, Q, R, P):
        '''
        Initialize the Kalman Filter.
        :param A: State transition matrix.
        :param B: Control input matrix.
        :param H: Measurement matrix.
        :param Q: Process noise covariance.
        :param R: Measurement noise covariance.
        :param P: Estimate error covariance.
        '''
        self.A = A
        self.B = B
        self.H = H
        self.Q = Q
        self.R = R
        self.P = P
```

```python
        self.x = np.zeros((A.shape[1], 1))

    def predict(self, u):
        '''
        Predict the state and estimate covariance.
        :param u: Control input.
        :return: Predicted state.
        '''
        self.x = np.dot(self.A, self.x) + np.dot(self.B, u)
        self.P = np.dot(np.dot(self.A, self.P), self.A.T) + self.Q
        return self.x

    def update(self, z):
        '''
        Update the state with the new measurement.
        :param z: Measurement input.
        :return: Updated state.
        '''
        y = z - np.dot(self.H, self.x)
        S = np.dot(self.H, np.dot(self.P, self.H.T)) + self.R
        K = np.dot(np.dot(self.P, self.H.T), np.linalg.inv(S))
        self.x = self.x + np.dot(K, y)
        I = np.eye(self.A.shape[0])
        self.P = np.dot(I - np.dot(K, self.H), self.P)
        return self.x

class RobustPIDController:
    def __init__(self, kp, ki, kd):
        '''
        Initialize the PID Controller.
        :param kp: Proportional gain.
        :param ki: Integral gain.
        :param kd: Derivative gain.
        '''
        self.kp = kp
        self.ki = ki
        self.kd = kd
        self.integral = 0
        self.previous_error = 0

    def compute_control(self, error, dt):
        '''
        Compute the control input using PID logic.
        :param error: Current error.
        :param dt: Time step.
        :return: Control input.
        '''
        self.integral += error * dt
        derivative = (error - self.previous_error) / dt
        control = self.kp * error + self.ki * self.integral +
        ↪    self.kd * derivative
        self.previous_error = error
        return control
```

245

```python
# Define matrices for Kalman Filter
A = np.array([[1, 0], [0, 1]])
B = np.array([[0.5], [1.0]])
H = np.array([[1, 0]])
Q = np.array([[0.01, 0], [0, 0.01]])
R = np.array([[0.1]])
P = np.array([[1, 0], [0, 1]])

# Initialize Kalman Filter and Robust PID Controller
kf = KalmanFilter(A, B, H, Q, R, P)
pid = RobustPIDController(kp=1.0, ki=0.1, kd=0.05)

# Simulate control scenario
control_input = []
state_estimates = []
time_steps = np.linspace(0, 10, 100)
measured_state = np.sin(time_steps)  # Simulated measurement

for t in range(len(time_steps)):
    current_measurement = measured_state[t]
    estimated_state = kf.predict(np.array([[0]]))
    updated_state = kf.update(np.array([[current_measurement]]))

    error = 1.0 - updated_state[0, 0]  # Desired state setpoint is
    ↪    1.0
    control_signal = pid.compute_control(error, 0.1)
    control_input.append(control_signal)
    state_estimates.append(updated_state[0, 0])

print("Control Inputs:", control_input)
print("State Estimates:", state_estimates)
```

This code defines critical functions necessary for the implementation and evaluation of robust GNC algorithms:

- KalmanFilter class implements a basic Kalman filter for state estimation, accommodating process and measurement noise.

- RobustPIDController class provides a framework for PID control that adjusts control inputs based on the error between desired and estimated states.

- The final block demonstrates simulating a control scenario using a KalmanFilter for state estimation and a RobustPIDController for control signal computation, equipped to handle dynamics under uncertainties.

This comprehensive system showcases the integration of estimation and control strategies crucial for aerospace applications, particularly under conditions where model uncertainties and disturbances impact performance.

Chapter 81

Computational Aeroelastic Optimization Algorithms

Below is a Python code snippet that demonstrates the implementation of an optimization algorithm considering aeroelastic effects in the design process of aerospace structures. This example uses a simplified model incorporating finite element analysis and optimization techniques to adjust structural parameters for aeroelastic performance.

```python
import numpy as np
from scipy.optimize import minimize
from scipy.linalg import eigh

# Define the stiffness and mass matrices (example values)
K = np.array([[12, -4, -2], [-4, 9, -3], [-2, -3, 7]])  # Stiffness
↪ matrix
M = np.array([[1, 0, 0], [0, 1, 0], [0, 0, 1]])         # Mass
↪ matrix

def aeroelastic_effect(eigenvalue, speed):
    # Computes simplified aeroelastic effect as a function of speed
    return 0.1 * eigenvalue * (speed ** 2)

def objective_function(params):
    speed, design_param1, design_param2 = params
```

```
# Adjust stiffness matrix with design parameters
K_adjusted = K + np.diag([design_param1, design_param2, 0])

# Solve generalized eigenvalue problem for natural frequencies
eigenvalues, _ = eigh(K_adjusted, M)

# Incorporate aeroelastic effects in the objective
obj = 0
for val in eigenvalues:
    obj += abs(val - aeroelastic_effect(val, speed))

return obj

# Initial parameters [speed, design_param1, design_param2]
initial_params = [100.0, 0.1, 0.1]

# Constraints and bounds for optimization
constraints = ({'type': 'ineq', 'fun': lambda x: x[0] - 50}) # speed
↪ > 50
bounds = [(50, 200), (-5, 5), (-5, 5)]                        #
↪ limits on design params

# Run optimization
result = minimize(objective_function, initial_params,
↪ constraints=constraints, bounds=bounds)

# Display result
print("Optimized Parameters:")
print("Flight Speed:", result.x[0])
print("Design Parameter 1:", result.x[1])
print("Design Parameter 2:", result.x[2])
```

This code defines an optimization process for an aeroelastic system:

- The `aeroelastic_effect` function models the impact of flight speed on aeroelastic properties through a simple relation.

- The `objective_function` computes an objective based on the deviation of eigenvalues from their aeroelastic-modified values, where modifications are functions of the design parameters and flight speed.

- The optimization minimizes the deviation across eigenvalues, thus tailoring structural design parameters to optimize for aeroelastic performance.

- The `minimize` function from `scipy.optimize` is employed to find optimal parameter values, under constraints and bounds.

By executing this optimization, the algorithm provides a conceptual framework for incorporating aeroelastic considerations directly into the design process, highlighting potential parameter adaptations that enhance performance. The example presented is simplified for demonstration purposes and may be expanded for more detailed aeroelastic models and realistic aerospace structures.

Chapter 82

Algorithms for Real-Time Simulation and Hardware-in-the-Loop Testing

Below is a Python code snippet that encompasses the core computational elements of real-time simulation in hardware-in-the-loop (HIL) environments, including real-time system data acquisition, control law implementation, and communication with external systems.

```python
import numpy as np
import time

class RealTimeController:
    def __init__(self, system_model, control_law):
        '''
        Initialize the real-time controller with a system model and
        ↪  control law.
        :param system_model: Function representing the system
        ↪  behavior.
        :param control_law: Function implementing the control law.
        '''
        self.system_model = system_model
        self.control_law = control_law
```

```python
        self.state = np.zeros(system_model.state_dimension())

    def acquire_data(self):
        '''
        Simulate data acquisition from sensors.
        :return: Current state of the system.
        '''
        # Simulate sensor data capture
        return self.system_model.current_state()

    def update_control(self, state):
        '''
        Compute the control action based on the current state.
        :param state: Current state of the system.
        :return: Control action.
        '''
        return self.control_law(state)

    def execute(self):
        '''
        Main execution loop for real-time control.
        '''
        while True:
            # Acquire current state
            self.state = self.acquire_data()

            # Compute control action
            control_action = self.update_control(self.state)

            # Apply control action to the system
            self.system_model.apply_control(control_action)

            # Simulate real-time operation by sleeping for a short
            ↪  duration
            time.sleep(0.01)

class SystemModel:
    def __init__(self, state_dimension):
        '''
        Initialize the system model with a given state dimension.
        :param state_dimension: Dimension of the state space.
        '''
        self._state = np.random.rand(state_dimension)

    def state_dimension(self):
        '''
        Get the dimension of the system state.
        :return: State dimension.
        '''
        return len(self._state)

    def current_state(self):
        '''
```

```
        Return the current state of the system.
        :return: Current state vector.
        '''
        return self._state

    def apply_control(self, control):
        '''
        Apply the control action to the system, updating its state.
        :param control: Control action to apply.
        '''
        self._state += control

def simple_control_law(state):
    '''
    Simple control law that attempts to stabilize the system.
    :param state: Current state of the system.
    :return: Control action.
    '''
    return -0.1 * state

# Example usage
if __name__ == '__main__':
    # Create system model and controller
    system = SystemModel(state_dimension=3)
    controller = RealTimeController(system_model=system,
    ↪    control_law=simple_control_law)

    # Start real-time control loop
    controller.execute()
```

This code defines several key components for real-time simulation in hardware-in-the-loop environments:

- The `RealTimeController` class manages the real-time loop, including data acquisition and control law execution.

- The `acquire_data` method simulates the process of collecting sensor data in real-time.

- The `update_control` method computes the control actions based on the acquired system state.

- The `SystemModel` class provides a simple dynamic model of the system being controlled.

- The `simple_control_law` function implements a basic feedback control law to stabilize the system.

This setup demonstrates the fundamental loop and components needed for real-time HIL simulation and control.

253

Chapter 83

Advanced Sampling Algorithms for Monte Carlo Simulations

Below is a Python code snippet that showcases an advanced sampling technique to improve the efficiency of Monte Carlo simulations in aerospace applications by incorporating stratified sampling.

```python
import numpy as np

def stratified_sample(n, bounds):
    '''
    Generates stratified random samples within given bounds.
    :param n: Number of strata (samples) to create.
    :param bounds: Tuple specifying lower and upper bounds (min,
    ↪    max).
    :return: Array of stratified samples.
    '''
    min_bound, max_bound = bounds
    strata_width = (max_bound - min_bound) / n
    # Generate stratified samples
    samples = np.array([np.random.uniform(min_bound + i *
    ↪    strata_width, min_bound + (i + 1) * strata_width) for i in
    ↪    range(n)])
    return samples

def function_to_integrate(x):
    '''
    A mock function to integrate using Monte Carlo methods.
    :param x: Input variable.
    :return: Function value at x.
```

```
    '''
    # Example function: the standard normal distribution probability
    ↪ density function
    return (1.0 / np.sqrt(2 * np.pi)) * np.exp(-0.5 * x**2)

def monte_carlo_integration_stratified(n, bounds):
    '''
    Perform stratified Monte Carlo integration over a specified
    ↪ interval.
    :param n: Number of samples/strata.
    :param bounds: Tuple specifying the lower and upper bounds (min,
    ↪ max).
    :return: Approximated integral value.
    '''
    samples = stratified_sample(n, bounds)
    samples_eval = function_to_integrate(samples)
    approx_integral = ((bounds[1] - bounds[0]) / n) *
    ↪ np.sum(samples_eval)
    return approx_integral

# Parameters for simulation
num_samples = 1000
integration_bounds = (-3, 3)

# Perform stratified Monte Carlo integration
approximate_value = monte_carlo_integration_stratified(num_samples,
↪ integration_bounds)
print(f"Approximated Integral Value: {approximate_value}")
```

This code provides an implementation of stratified sampling to
improve Monte Carlo simulation efficiency:

- `stratified_sample` function generates stratified random samples within specified bounds to ensure even coverage over the
  sample space.

- `function_to_integrate` represents a mock function to be
  integrated using Monte Carlo methods, such as a probability
  density function.

- `monte_carlo_integration_stratified` performs integration
  by evaluating the function over stratified samples, providing
  an efficient way to approximate integrals.

The final portion of the code demonstrates the application of
these functions to perform Monte Carlo integration with stratified
sampling, achieving a more accurate result with fewer samples than
simple random sampling would require.

Chapter 84

Computational Modeling of Jet Noise Reduction

Below is a Python code snippet that encompasses the core computational elements for simulating and mitigating noise generated by jet engines. This includes the modeling of jet noise sources, applying noise mitigation strategies, and analyzing the impact of design changes on noise levels.

```python
import numpy as np
import matplotlib.pyplot as plt

def calculate_jet_noise(frequency, velocity, diameter):
    '''
    Calculate jet noise using a simplified version of Lighthill's
    ↪ Acoustic Analogy.
    :param frequency: Frequency of the noise.
    :param velocity: Jet velocity.
    :param diameter: Jet diameter.
    :return: Noise intensity level.
    '''
    rho = 1.225  # Air density in kg/m^3
    c = 343  # Speed of sound in m/s
    lighthill_constant = (8 / 3) * (rho**2) * (velocity**8) / (c**5
    ↪ * diameter**2)

    intensity = lighthill_constant * (frequency**2)
    return intensity
```

```
def apply_noise_reduction_technology(noise_levels,
↪   reduction_factor):
    '''
    Apply noise reduction technology to reduce noise levels.
    :param noise_levels: Array of calculated noise levels.
    :param reduction_factor: Factor by which noise is mitigated
    ↪   (e.g., 0.1 for 10% reduction).
    :return: Mitigated noise levels.
    '''

    return noise_levels * (1 - reduction_factor)

def plot_noise_spectrum(frequencies, noise_levels,
↪   mitigated_noise_levels):
    '''
    Plot the noise spectrum before and after noise reduction.
    :param frequencies: Array of noise frequencies.
    :param noise_levels: Original noise levels.
    :param mitigated_noise_levels: Mitigated noise levels.
    '''
    plt.figure(figsize=(10, 6))
    plt.plot(frequencies, noise_levels, label='Original Noise')
    plt.plot(frequencies, mitigated_noise_levels, label='Mitigated
    ↪   Noise', linestyle='--')
    plt.xlabel('Frequency (Hz)')
    plt.ylabel('Noise Level (Arbitrary Units)')
    plt.title('Jet Noise Spectrum')
    plt.legend()
    plt.grid(True)
    plt.show()

# Example simulation parameters
frequencies = np.linspace(100, 5000, 100)  # Frequency range from
↪   100 Hz to 5000 Hz
velocity = 300  # Jet velocity in m/s
diameter = 0.5  # Jet diameter in meters

# Calculate original jet noise levels
original_noise_levels = np.array([calculate_jet_noise(f, velocity,
↪   diameter) for f in frequencies])

# Apply noise reduction technology
reduction_factor = 0.15  # Reduce noise by 15%
mitigated_noise_levels =
↪   apply_noise_reduction_technology(original_noise_levels,
↪   reduction_factor)

# Plot the noise spectrum
plot_noise_spectrum(frequencies, original_noise_levels,
↪   mitigated_noise_levels)
```

This code defines several key functions necessary for the simulation and mitigation of jet engine noise:

- `calculate_jet_noise` function computes the noise intensity generated by a jet based on the frequency, velocity, and diameter using a simplified Lighthill's Acoustic Analogy.

- `apply_noise_reduction_technology` reduces calculated noise levels by a specified factor, simulating the effects of noise mitigation technologies.

- `plot_noise_spectrum` visualizes the original and mitigated noise spectra, aiding in the analysis of noise reduction impacts.

The final block of code provides a demonstration of these functions with example parameters for a typical jet, illustrating noise calculation and reduction visually.

Chapter 85

Algorithms for Electric and Hybrid Propulsion Systems

Below is a Python code snippet that demonstrates the simulation of electric and hybrid propulsion systems, focusing on basic propulsion modeling, energy consumption, and simple control strategies for performance optimization.

```python
import numpy as np

def thrust_force(power, efficiency, velocity):
    '''
    Calculate the thrust force generated by the propulsion system.
    :param power: The power input to the propulsion system (in
    ↪    watts).
    :param efficiency: Efficiency of the propulsion system (0 to 1).
    :param velocity: Velocity of the vehicle (in m/s).
    :return: Generated thrust force in Newtons (N).
    '''
    if velocity == 0:
        return 0
    return (power * efficiency) / velocity

def energy_consumption(thrust, time, efficiency):
    '''
    Calculate the energy consumption of the propulsion system.
    :param thrust: Thrust force generated (in N).
    :param time: Operational time (in seconds).
    :param efficiency: Efficiency of the propulsion system (0 to 1).
    :return: Total energy consumed (in joules).
```

```
    '''
    return (thrust * velocity * time) / efficiency

def hybrid_control_strategy(battery_level, required_thrust,
↪    max_motor_power, fuel_efficiency):
    '''
    Determine the distribution of power between electric and fuel
    ↪    sources.
    :param battery_level: Current battery level (as a percentage).
    :param required_thrust: Required thrust to maintain operation
    ↪    (in N).
    :param max_motor_power: Maximum power the electric motor can
    ↪    supply (in watts).
    :param fuel_efficiency: Efficiency of fuel usage.
    :return: Distribution plan (power from battery, power from
    ↪    fuel).
    '''

    electric_power = min(max_motor_power, battery_level *
    ↪    max_motor_power / 100)
    remaining_thrust = required_thrust -
    ↪    thrust_force(electric_power, 0.85, velocity)

    if remaining_thrust > 0:
        fuel_power = (remaining_thrust * velocity) / fuel_efficiency
    else:
        fuel_power = 0

    return electric_power, fuel_power

# Example scenario parameters
power_input = 50000  # in watts
efficiency = 0.85  # 85%
velocity = 200  # in m/s
battery_level = 75  # 75%
required_thrust = 1600  # in N
max_motor_power = 30000  # in watts
fuel_efficiency = 0.3  # 30%

# Simulation
thrust = thrust_force(power_input, efficiency, velocity)
energy_used = energy_consumption(thrust, 3600, efficiency)
electric_power, fuel_power = hybrid_control_strategy(battery_level,
↪    required_thrust, max_motor_power, fuel_efficiency)

print("Thrust Force:", thrust, "N")
print("Total Energy Used:", energy_used, "Joules")
print("Electric Power Used:", electric_power, "Watts")
print("Fuel Power Required:", fuel_power, "Watts")
```

This code defines several key functions necessary for the modeling and control of hybrid and electric propulsion systems:

- `thrust_force` function computes the thrust force based on power input, system efficiency, and current velocity.

- `energy_consumption` estimates the energy consumption required to sustain the propulsion system's operation for a given time.

- `hybrid_control_strategy` determines the balance between battery power and fuel needs to achieve required thrust, optimizing the use of available resources.

The final segment of the code provides an example scenario with predefined parameters, simulating the interaction of the propulsion system's components.

Chapter 86

Computational Nanotechnology in Aerospace Engineering

Below is a Python code snippet that encompasses the core computational elements for simulating nanomaterials and their applications in aerospace systems. It includes a molecular dynamics simulation for a simplified nanomaterial system, integration with aerospace component models, and evaluation of material properties relevant to aerospace applications.

```python
import numpy as np
from scipy.constants import k, atomic_mass, elementary_charge

def lennard_jones_potential(r, epsilon, sigma):
    '''
    Calculate the Lennard-Jones potential between a pair of
    ↪  particles.
    :param r: Distance between particles.
    :param epsilon: Depth of the potential well.
    :param sigma: Distance at which the potential is zero.
    :return: Potential energy.
    '''
    return 4 * epsilon * ((sigma / r)**12 - (sigma / r)**6)

def initialize_positions(num_particles, box_size):
    '''
    Initializes positions of particles within a cubic box.
    :param num_particles: Number of particles.
    :param box_size: Size of the simulation box.
```

```python
    :return: Array of particle positions.
    '''
    return np.random.rand(num_particles, 3) * box_size

def initialize_velocities(num_particles, temperature, mass):
    '''
    Initializes velocities based on a Maxwell-Boltzmann
    ↪ distribution.
    :param num_particles: Number of particles.
    :param temperature: Temperature in Kelvin.
    :param mass: Mass of each particle.
    :return: Array of particle velocities.
    '''
    velocities = np.random.randn(num_particles, 3) * np.sqrt(k *
    ↪ temperature / mass)
    velocities -= np.mean(velocities, axis=0)    # Remove
    ↪ center-of-mass motion
    return velocities

def integrate_motion(positions, velocities, forces, dt, mass):
    '''
    Integrates the equations of motion for the particles.
    :param positions: Current positions of particles.
    :param velocities: Current velocities of particles.
    :param forces: Current forces on particles.
    :param dt: Time step.
    :param mass: Mass of each particle.
    :return: Updated positions and velocities.
    '''
    new_positions = positions + velocities * dt + 0.5 * forces *
    ↪ dt**2 / mass
    new_velocities = velocities + forces * dt / mass
    return new_positions, new_velocities

def compute_forces(positions, epsilon, sigma, cutoff):
    '''
    Computes forces on each particle due to the Lennard-Jones
    ↪ potential.
    :param positions: Current positions of particles.
    :param epsilon: Lennard-Jones potential depth.
    :param sigma: Distance at which potential is zero.
    :param cutoff: Distance beyond which interactions are not
    ↪ considered.
    :return: Array of forces on particles.
    '''
    num_particles = positions.shape[0]
    forces = np.zeros_like(positions)
    for i in range(num_particles):
        for j in range(i + 1, num_particles):
            rij = positions[i] - positions[j]
            r = np.linalg.norm(rij)
            if r < cutoff:
```

```
            f = (24 * epsilon / r) * (2 * (sigma / r)**12 -
            ↪ (sigma / r)**6) * rij / r
            forces[i] += f
            forces[j] -= f
    return forces

def simulate_md(num_steps, dt, num_particles, box_size, temperature,
↪ mass, epsilon, sigma, cutoff):
    ' ' '
    Performs a molecular dynamics simulation.
    :param num_steps: Number of simulation steps.
    :param dt: Time step.
    :param num_particles: Number of particles.
    :param box_size: Size of simulation box.
    :param temperature: Temperature of the system.
    :param mass: Mass of each particle.
    :param epsilon: Lennard-Jones potential depth.
    :param sigma: Lennard-Jones interaction distance.
    :param cutoff: Interaction cutoff distance.
    :return: Final positions and velocities of particles.
    ' ' '
    positions = initialize_positions(num_particles, box_size)
    velocities = initialize_velocities(num_particles, temperature,
    ↪ mass)
    for step in range(num_steps):
        forces = compute_forces(positions, epsilon, sigma, cutoff)
        positions, velocities = integrate_motion(positions,
        ↪ velocities, forces, dt, mass)
    return positions, velocities

# Simulate properties of a nanomaterial system
final_positions, final_velocities = simulate_md(
    num_steps=1000,
    dt=1e-3,
    num_particles=100,
    box_size=10.0,
    temperature=300,
    mass=40.0 * atomic_mass,
    epsilon=0.0103 * elementary_charge,
    sigma=3.4,
    cutoff=8.5
)

# Analyze and apply these properties in aerospace contexts here
# Placeholder for further application-specific analysis

print("Final positions:", final_positions)
print("Final velocities:", final_velocities)
```

The provided code establishes a molecular dynamics framework
for simulating nanomaterials, crucial for understanding their performance in aerospace applications:

264

- `lennard_jones_potential` computes the interaction energy between particles using the Lennard-Jones potential.

- `initialize_positions` and `initialize_velocities` set up the initial state of the system reflecting a random distribution akin to physical conditions.

- `integrate_motion` moves the simulation forward in time, adjusting particle positions and velocities.

- `compute_forces` calculates forces considering particle interactions within designated cutoff distances.

- `simulate_md` runs the core molecular dynamics process, evolving the system over multiple time steps.

This simulation code lays the foundation for deeper exploration into the integration of nanomaterials in aerospace systems, such as assessing durability, thermal resistance, and mechanical properties.

Chapter 87

Algorithms for Hypersonic Boundary Layer Transition

Below is a Python code snippet that demonstrates a simulation for predicting the transition from laminar to turbulent flow in hypersonic boundary layers. This includes the implementation of various computational techniques such as stability analysis and transition prediction models.

```python
import numpy as np
import matplotlib.pyplot as plt

def hypersonic_boundary_layer(x, U_inf, rho_inf, mu_inf, Pr, T_inf,
↪ gamma, R):
    '''
    Compute boundary layer properties for hypersonic flow.
    :param x: Spatial coordinate.
    :param U_inf: Freestream velocity.
    :param rho_inf: Freestream density.
    :param mu_inf: Freestream viscosity.
    :param Pr: Prandtl number.
    :param T_inf: Freestream temperature.
    :param gamma: Specific heat ratio.
    :param R: Gas constant.
    :return: Velocity and temperature profile, and Re_theta.
    '''
    T_aw = T_inf * (1 + (gamma-1)/2 * (U_inf/R/T_inf)**2)
    T_w = T_inf  # Adiabatic wall assumption
    nu_inf = mu_inf / rho_inf
```

```python
    Re_theta = U_inf * x / nu_inf

    y = np.linspace(0, 0.01, num=100)  # Non-dimensional distance
    u_profile = U_inf * np.tanh(y * Re_theta**0.5)
    T_profile = T_w + (T_aw - T_w) * (1 -
    ↪  np.exp(-Pr*y*Re_theta**0.5))

    return u_profile, T_profile, Re_theta

def e_n_method(Rex, n=9):
    '''
    e^n method for transition prediction.
    :param Rex: Reynolds number based on x.
    :param n: Critical amplification factor.
    :return: Transition prediction.
    '''

    return Rex >= np.exp(n)

# Example parameters for a hypersonic flow
x = 0.05   # 5 cm from the leading edge
U_inf = 1500   # m/s
rho_inf = 0.018   # kg/m^3 (roughly at 30 km altitude)
mu_inf = 1.46e-5   # kg/(m*s)
Pr = 0.72
T_inf = 226.5   # Kelvin
gamma = 1.4
R = 287.05   # J/(kg*K)

# Compute boundary layer properties
u_profile, T_profile, Re_theta = hypersonic_boundary_layer(x, U_inf,
↪  rho_inf, mu_inf, Pr, T_inf, gamma, R)
# Determine transition using e^n method
transition_occurred = e_n_method(Re_theta)

print("Transition occurred:", transition_occurred)
print("Re_theta:", Re_theta)

# Plotting velocity and temperature profiles
plt.figure()
plt.subplot(2, 1, 1)
plt.plot(u_profile, label='Velocity Profile')
plt.xlabel('Non-dimensional Distance')
plt.ylabel('Velocity (m/s)')
plt.title('Velocity Profile of Hypersonic Boundary Layer')
plt.grid()
plt.legend()

plt.subplot(2, 1, 2)
plt.plot(T_profile, label='Temperature Profile', color='r')
plt.xlabel('Non-dimensional Distance')
plt.ylabel('Temperature (K)')
plt.title('Temperature Profile of Hypersonic Boundary Layer')
plt.grid()
```

```
plt.legend()

plt.tight_layout()
plt.show()
```

This code defines functions necessary for predicting the transition from laminar to turbulent flow using hypersonic boundary layer calculations:

- `hypersonic_boundary_layer` computes the velocity and temperature profiles across the boundary layer and calculates the momentum thickness Reynolds number (`Re_theta`).

- `e_n_method` implements the e^n method for transition prediction, checking if the `Re_theta` exceeds an exponential factor indicating likely transition.

These methods are exemplified with parameters typical for a hypersonic flow scenario, with output plots illustrating the velocity and temperature profiles.

Chapter 88

Computational Methods for Autonomous Rendezvous and Docking

Below is a Python code snippet that encompasses the core computational elements of algorithms enabling spacecraft to autonomously rendezvous and dock with other vehicles or stations. The code includes the implementation of rendezvous trajectory generation, control strategies, and sensor data handling.

```python
import numpy as np
from scipy.integrate import solve_ivp

def rendezvous_dynamics(t, state, target_position, control_law):
    '''
    Computes the dynamics of the chaser spacecraft for rendezvous.
    :param t: Current time in the simulation.
    :param state: Current state vector [x, y, z, vx, vy, vz] of the
    ↪ chaser.
    :param target_position: Target spacecraft position [x, y, z].
    :param control_law: Function to compute control acceleration.
    :return: Derivative of the state vector.
    '''

    position = state[:3]
```

```
    velocity = state[3:]

    # Calculate control input
    control_input = control_law(position, velocity, target_position)

    # Simple linear model for dynamics
    dxdt = velocity
    dvdt = control_input

    return np.concatenate((dxdt, dvdt))

def proportional_navigation_guidance(position, velocity,
↪   target_position, gain=1.0):
    '''
    Proportional navigation guidance law for rendezvous.
    :param position: Current position of the chaser [x, y, z].
    :param velocity: Current velocity of the chaser [vx, vy, vz].
    :param target_position: Target spacecraft position [x, y, z].
    :param gain: Gain for the proportional navigation.
    :return: Control acceleration [ax, ay, az].
    '''

    relative_position = target_position - position
    relative_velocity = -velocity

    los_rate = np.cross(relative_position, relative_velocity) /
↪   np.dot(relative_position, relative_position)
    control_acceleration = gain * np.cross(relative_velocity,
↪   los_rate)

    return control_acceleration

def simulate_rendezvous(initial_state, target_position, duration):
    '''
    Simulates the rendezvous maneuver using an ODE solver.
    :param initial_state: Initial state vector [x, y, z, vx, vy, vz]
↪   of the chaser.
    :param target_position: Target spacecraft position [x, y, z].
    :param duration: Duration of the simulation.
    :return: Solution object containing the state trajectory.
    '''

    control_law = lambda pos, vel, target:
↪   proportional_navigation_guidance(pos, vel, target)
    sol = solve_ivp(rendezvous_dynamics, [0, duration],
↪   initial_state, args=(target_position, control_law),
            t_eval=np.linspace(0, duration, num=1000))
    return sol

# Example initial state and target position
initial_state = np.array([1000, 0, 0, 0, 1, 0])  # Position [x, y,
↪   z] and velocity [vx, vy, vz]
target_position = np.array([0, 0, 0])  # Target at origin

# Simulate the rendezvous
```

```
trajectory = simulate_rendezvous(initial_state, target_position,
↪   duration=5000)

# Output the final position and velocity to check the rendezvous
↪   outcome
final_state = trajectory.y[:, -1]
print("Final Position:", final_state[:3], "Final Velocity:",
↪   final_state[3:])
```

This code defines several key functions necessary for the implementation of spacecraft rendezvous and docking algorithms:

- `rendezvous_dynamics` function computes the dynamics of the chaser spacecraft, providing updates to its position and velocity based on control inputs.

- `proportional_navigation_guidance` implements a guidance policy that adjusts the chaser's trajectory to align with the target using a proportional navigation technique.

- `simulate_rendezvous` utilizes an ODE solver to simulate the chaser's path as it moves towards the target position over a specified duration.

The final block of code provides a demonstration of these elements with an example initial state and a target at the origin, simulating the rendezvous process and printing the final state.

Chapter 89

Algorithms for Active Flow Control

Below is a Python code snippet that encompasses the core computational elements of implementing active flow control strategies to manipulate aerodynamic flows, focusing on the use of synthetic jets as a control mechanism.

```python
import numpy as np
import matplotlib.pyplot as plt

def compute_synthetic_jet_velocity(amplitude, frequency, time):
    '''
    Calculate the velocity of a synthetic jet at a given time.
    :param amplitude: The amplitude of the jet oscillation.
    :param frequency: The frequency of the jet oscillation.
    :param time: The time at which to calculate the velocity.
    :return: Instantaneous velocity of the synthetic jet.
    '''
    return amplitude * np.sin(2 * np.pi * frequency * time)

def control_flow_field(u_inf, jet_amplitude, jet_frequency,
↪ time_array):
    '''
    Apply synthetic jet flow control to the main flow field.
    :param u_inf: Freestream velocity.
    :param jet_amplitude: Amplitude of the jet oscillation.
    :param jet_frequency: Frequency of the jet oscillation.
    :param time_array: Array of time points for which the control is
    ↪ applied.
    :return: Modified flow field with the control applied.
    '''
    # Initializing flow field
```

```
velocity_field = np.zeros_like(time_array)

for i, t in enumerate(time_array):
    jet_velocity = compute_synthetic_jet_velocity(jet_amplitude,
    ↪  jet_frequency, t)
    velocity_field[i] = u_inf + jet_velocity

return velocity_field

def plot_velocity_field(time_array, velocity_field):
    '''
    Plot the velocity field over time.
    :param time_array: Array of time points.
    :param velocity_field: The velocity field to plot.
    '''
    plt.figure()
    plt.plot(time_array, velocity_field, label='Controlled Flow')
    plt.xlabel('Time (s)')
    plt.ylabel('Velocity (m/s)')
    plt.title('Velocity Field with Active Flow Control')
    plt.legend()
    plt.grid(True)
    plt.show()

# Parameters for synthetic jet and flow field
u_inf = 10.0  # Freestream velocity in m/s
jet_amplitude = 1.0  # Amplitude of synthetic jet in m/s
jet_frequency = 1.0  # Frequency in Hz
time_array = np.linspace(0, 5, 500)  # Time array from 0 to 5
↪  seconds

# Compute the modified flow field
velocity_field = control_flow_field(u_inf, jet_amplitude,
↪  jet_frequency, time_array)

# Visualize the results
plot_velocity_field(time_array, velocity_field)
```

This code defines several key functions necessary for implementing active flow control using synthetic jets:

- compute_synthetic_jet_velocity computes the instantaneous velocity of a synthetic jet driven by specified amplitude and frequency.

- control_flow_field applies the jet control to the main flow field, superimposing the synthetic jet velocity onto the freestream velocity.

- plot_velocity_field provides a visualization of the velocity field over time, illustrating the effects of the active flow control.

The final block of code demonstrates how to set up and execute these functions, producing a plot of the controlled flow field dynamics over time.

Chapter 90

Computational Methods for Thermal Protection Systems

Below is a Python code snippet that encompasses the core computational elements of algorithms for designing and analyzing materials and structures that protect against extreme thermal environments.

```python
import numpy as np

class ThermalProtectionSystem:
    def __init__(self, material_properties, boundary_conditions):
        '''
        Initialize the thermal protection system with material
        ↪  properties and boundary conditions.
        :param material_properties: Dictionary containing thermal
        ↪  conductivity, specific heat, density.
        :param boundary_conditions: Dictionary for initial and
        ↪  boundary conditions of the system.
        '''
        self.k = material_properties['conductivity']
        self.c = material_properties['specific_heat']
        self.rho = material_properties['density']
        self.T_initial = boundary_conditions['initial_temperature']
        self.T_boundary =
        ↪  boundary_conditions['boundary_temperature']
        self.dx = boundary_conditions['spatial_step']
        self.dt = boundary_conditions['time_step']
        self.length = boundary_conditions['length']
        self.grid_points = int(self.length / self.dx)
```

```python
        self.temperature_field = np.ones(self.grid_points) *
        ↪   self.T_initial

    def compute_temperature_distribution(self, time):
        '''
        Run the simulation over the specified time to compute
        ↪   temperature distribution.
        :param time: Total time for which to run the simulation.
        '''
        alpha = self.k / (self.c * self.rho)  # Thermal diffusivity
        time_steps = int(time / self.dt)

        for step in range(time_steps):
            new_temperature_field = np.copy(self.temperature_field)
            for i in range(1, self.grid_points - 1):
                new_temperature_field[i] = self.temperature_field[i]
                ↪   + alpha * self.dt / self.dx**2 * (
                    self.temperature_field[i + 1] - 2 *
                    ↪   self.temperature_field[i] +
                    ↪   self.temperature_field[i - 1])

            # Apply boundary conditions
            new_temperature_field[0] = self.T_boundary
            new_temperature_field[-1] = self.T_boundary

            self.temperature_field = new_temperature_field

    def get_temperature_profile(self):
        '''
        Get the current temperature profile of the system.
        :return: Array of temperature values across the material.
        '''
        return self.temperature_field

# Example material properties and boundary conditions
material_props = {
    'conductivity': 0.5,      # Thermal conductivity in W/(m*K)
    'specific_heat': 900,     # Specific heat in J/(kg*K)
    'density': 2700           # Density in kg/m^3
}

boundary_conditions = {
    'initial_temperature': 300,    # Initial temperature in Kelvin
    'boundary_temperature': 1000,  # Boundary temperature in Kelvin
    'spatial_step': 0.01,          # Spatial step in meters
    'time_step': 0.1,              # Time step in seconds
    'length': 1.0                  # Length of the material in
    ↪   meters
}

# Initialize and run the thermal simulation
tps = ThermalProtectionSystem(material_props, boundary_conditions)
```

276

```
tps.compute_temperature_distribution(time=100)  # Simulate for 100
↪    seconds

print("Final Temperature Distribution:",
↪    tps.get_temperature_profile())
```

This code defines several key classes and methods necessary for simulating a thermal protection system:

- **ThermalProtectionSystem** class initializes the system with given material properties and boundary conditions, setting up the spatial and temporal grid.

- **compute_temperature_distribution** method performs the simulation over a specified time to compute how temperature propagates through the material.

- **get_temperature_profile** method returns the current temperature profile across the material after simulation.

The provided example demonstrates how to apply these classes and methods with specific material properties and boundary conditions to simulate a thermal protection scenario.

Chapter 91

Algorithms for Space Weather Prediction

Below is a Python code snippet that illustrates a basic computational model for predicting solar activity and its potential impact on aerospace operations. This includes fetching space weather data, processing it to predict solar events, and calculating the potential effects on aerospace systems.

```python
import numpy as np
import pandas as pd
from sklearn.linear_model import LinearRegression
from sklearn.preprocessing import StandardScaler
from scipy.signal import find_peaks

def fetch_solar_data(file_path):
    '''
    Fetch solar activity data from a CSV file.
    :param file_path: Path to the CSV file containing solar activity
    ↪ data.
    :return: DataFrame containing the imported data.
    '''
    data = pd.read_csv(file_path)
    return data

def preprocess_data(data):
    '''
    Preprocess the data by standardizing and detecting relevant
    ↪ features.
    :param data: DataFrame containing the solar activity data.
    :return: Preprocessed data ready for predictive modeling.
    '''
```

```
    scaler = StandardScaler()
    scaled_data = scaler.fit_transform(data)
    return pd.DataFrame(scaled_data, columns=data.columns)

def predict_solar_activity(data):
    '''
    Predict future solar activity based on historical data using a
    ↪ simplistic linear regression model.
    :param data: Preprocessed historical solar activity data.
    :return: Predicted solar activity indices.
    '''
    model = LinearRegression()
    X = np.arange(len(data)).reshape(-1, 1)
    y = data['solar_index']
    model.fit(X, y)
    future_times = np.arange(len(data), len(data) + 30).reshape(-1,
    ↪ 1)
    predictions = model.predict(future_times)
    return predictions

def calculate_aerospace_impacts(solar_predictions, threshold=1.5):
    '''
    Calculate potential impacts on aerospace operations based on
    ↪ predicted solar activity.
    :param solar_predictions: Predicted solar activity indices.
    :param threshold: Impact threshold for solar activity.
    :return: Array indicating presence of potential impacts.
    '''
    impacts = np.zeros_like(solar_predictions)
    peaks, _ = find_peaks(solar_predictions, height=threshold)
    impacts[peaks] = 1
    return impacts

# Example usage
file_path = 'solar_activity.csv'  # Assume this file contains
↪ indexed solar activity data
data = fetch_solar_data(file_path)
preprocessed_data = preprocess_data(data)
solar_predictions = predict_solar_activity(preprocessed_data)
impacts = calculate_aerospace_impacts(solar_predictions)

print("Predicted Solar Activity:", solar_predictions)
print("Potential Aerospace Impacts:", impacts)
```

This code snippet includes several functions that demonstrate the prediction of solar activity and its impacts:

- `fetch_solar_data` imports solar activity data from a specified CSV file.

- `preprocess_data` standardizes the data for further analysis and modeling.

- `predict_solar_activity` predicts future solar activity using linear regression based on historical data.

- `calculate_aerospace_impacts` identifies potential impacts on aerospace operations if predicted solar activity exceeds a given threshold.

These functions collectively provide a baseline methodology for assessing how solar activity might affect aerospace systems, potentially guiding operational protocols and strategic decisions.

Chapter 92

Computational Analysis of High-Altitude Aerodynamics

Below is a Python code snippet that encompasses the core algorithms for simulating aerodynamic phenomena in near-space and exoatmospheric environments, specifically focusing on fluid dynamics using simplified models for gas dynamics, thermodynamics, and aerodynamic forces.

```python
import numpy as np

def atmospheric_density(altitude):
    '''
    Calculate atmospheric density based on altitude.
    :param altitude: Altitude in meters.
    :return: Density in kg/m^3.
    '''
    if altitude < 11000:
        # Troposphere
        return 1.225 * (1 - (altitude / 44330)) ** 4.256
    elif altitude < 20000:
        # Lower Stratosphere
        return 0.3639 * np.exp(-(altitude - 11000) / 6341.6)
    else:
        # Upper Layer (approximation)
        return 0
```

```python
def drag_force(C_D, rho, A, V):
    '''
    Calculate drag force.
    :param C_D: Drag coefficient.
    :param rho: Density of the fluid.
    :param A: Reference area.
    :param V: Velocity.
    :return: Drag force in Newtons.
    '''
    return 0.5 * C_D * rho * A * V**2

def lift_force(C_L, rho, A, V):
    '''
    Calculate lift force.
    :param C_L: Lift coefficient.
    :param rho: Density of the fluid.
    :param A: Reference area.
    :param V: Velocity.
    :return: Lift force in Newtons.
    '''
    return 0.5 * C_L * rho * A * V**2

def simulate_aerodynamics(altitude, velocity, C_L, C_D, A):
    '''
    Simulate aerodynamic forces at a given altitude and velocity.
    :param altitude: Altitude in meters.
    :param velocity: Velocity in m/s.
    :param C_L: Lift coefficient.
    :param C_D: Drag coefficient.
    :param A: Reference area.
    :return: Lift, drag forces.
    '''
    rho = atmospheric_density(altitude)
    lift = lift_force(C_L, rho, A, velocity)
    drag = drag_force(C_D, rho, A, velocity)
    return lift, drag

# Example parameters for simulation
altitude = 20000   # altitude in meters
velocity = 1500    # velocity in m/s
C_L = 0.5          # lift coefficient (hypothetical)
C_D = 0.3          # drag coefficient (hypothetical)
A = 10             # reference area in m^2

# Simulate forces
lift, drag = simulate_aerodynamics(altitude, velocity, C_L, C_D, A)

print("Lift Force:", lift, "N")
print("Drag Force:", drag, "N")
```

This code defines several key functions necessary for understanding aerodynamic phenomena in near-space environments:

- `atmospheric_density` function estimates the atmospheric density based on altitude, utilizing a simplistic model for different atmospheric layers.

- `drag_force` calculates the aerodynamic drag force acting on an object given its velocity, drag coefficient, and reference area.

- `lift_force` computes the lift force generated, which is crucial for sustaining flight in thin atmospheres.

- `simulate_aerodynamics` coordinates the application of these principles to simulate lift and drag forces at specific altitudes and velocities.

The final block of code provides examples of computing aerodynamic forces using defined parameters, illustrating the effects at high altitudes and speeds typical of near-space operations.

Chapter 93

Algorithms for Swarm Robotics in Aerospace

Below is a Python code snippet that encompasses the core computational elements of coordinating multiple robotic agents in aerospace applications, including path planning, communication, and task allocation for a simple swarm of UAVs.

```python
import numpy as np

class UAV:
    def __init__(self, id, position, communication_range):
        self.id = id
        self.position = np.array(position)
        self.communication_range = communication_range
        self.task = None

    def distance_to(self, other_uav):
        return np.linalg.norm(self.position - other_uav.position)

    def can_communicate_with(self, other_uav):
        return self.distance_to(other_uav) <=
        ↪   self.communication_range

    def assign_task(self, task_position):
        self.task = task_position
        print(f"UAV {self.id} assigned to task at position
        ↪   {task_position}")

    def move_towards_task(self, step_size=1.0):
        if self.task is not None:
            direction = self.task - self.position
            if np.linalg.norm(direction) < step_size:
```

```
                    self.position = self.task
                    self.task = None
                    print(f"UAV {self.id} has reached its task.")
                else:
                    direction = direction / np.linalg.norm(direction) *
                    ↪    step_size
                    self.position += direction
                    print(f"UAV {self.id} moved to {self.position}")

    def plan_paths(uav_swarm, tasks):
        for task_position in tasks:
            closest_uav = min(uav_swarm, key=lambda uav:
            ↪    np.linalg.norm(uav.position - task_position))
            closest_uav.assign_task(task_position)

    def simulate_swarm_operation(uav_swarm, steps=10):
        for step in range(steps):
            print(f"Step {step + 1}")
            for uav in uav_swarm:
                uav.move_towards_task()

    # Define UAV swarm
    uav_swarm = [
        UAV(id=1, position=[0, 0], communication_range=30),
        UAV(id=2, position=[10, 10], communication_range=30),
        UAV(id=3, position=[20, 5], communication_range=30)
    ]

    # Define task locations
    tasks = [
        np.array([15, 0]),
        np.array([25, 15]),
        np.array([5, 25])
    ]

    # Plan paths for UAVs to complete tasks
    plan_paths(uav_swarm, tasks)

    # Simulate the operation of the swarm, moving towards tasks
    simulate_swarm_operation(uav_swarm)
```

This code defines several core functionalities for coordinating a swarm of UAVs:

- UAV class represents individual UAVs with attributes for position and communication range, and methods for task assignment and movement.

- `distance_to` and `can_communicate_with` methods calculate distances between UAVs and determine possible communication pairs.

- `assign_task` assigns a task based on proximity, and `move_towards_task` simulates navigation towards the task.

- `plan_paths` is responsible for allocating tasks to UAVs based on their current positions.

- `simulate_swarm_operation` animates the process of moving UAVs towards their assigned tasks, showing step-by-step progress.

This Python code snippet exhibits fundamental mechanisms that can be further expanded to handle more sophisticated scenarios, such as dynamic task reassignment, collision avoidance, and energy-efficient path planning.

Chapter 94

Optimization Algorithms for Crew Scheduling and Maintenance

Below is a Python code snippet that implements core algorithms for optimizing crew assignments and maintenance schedules in aviation operations, focusing on formulating the optimization problem and solving it using mixed-integer linear programming.

```python
from ortools.linear_solver import pywraplp

def crew_assignment_optimization(crews, flights, maintenance_slots,
↪   flight_crew_requirements, crew_availability,
↪   maintenance_requirements):
    '''
    Optimize crew assignments and maintenance schedules.
    :param crews: List of crew members.
    :param flights: List of flights.
    :param maintenance_slots: Available maintenance slots.
    :param flight_crew_requirements: Dictionary of flight to crew
    ↪   requirements.
    :param crew_availability: Dictionary of crew availability per
    ↪   flight.
    :param maintenance_requirements: Dictionary of flights to
    ↪   maintenance needs.
    :return: Optimal crew and maintenance assignments.
    '''
```

```
solver = pywraplp.Solver.CreateSolver('SCIP')

# Decision variables
x = {}
for crew in crews:
    for flight in flights:
        x[crew, flight] = solver.IntVar(0, 1,
        ↪  f'x[{crew},{flight}]')

y = {}
for flight in flights:
    for slot in maintenance_slots:
        y[flight, slot] = solver.IntVar(0, 1,
        ↪  f'y[{flight},{slot}]')

# Constraints
for flight in flights:
    solver.Add(sum(x[crew, flight] for crew in crews) >=
    ↪  flight_crew_requirements[flight])

for crew in crews:
    solver.Add(sum(x[crew, flight] for flight in flights if
    ↪  crew_availability[crew][flight]) <= 1)

for flight in flights:
    solver.Add(sum(y[flight, slot] for slot in
    ↪  maintenance_slots) == maintenance_requirements[flight])

# Objective
solver.Maximize(
    sum(x[crew, flight] for crew in crews for flight in flights)
    ↪  +
    sum(y[flight, slot] for flight in flights for slot in
    ↪  maintenance_slots)
)

status = solver.Solve()

# Output Results
if status == pywraplp.Solver.OPTIMAL:
    print('Optimal solution found!')
    crew_assignment = {flight: [crew for crew in crews if
    ↪  x[crew, flight].solution_value() == 1] for flight in
    ↪  flights}
    maintenance_schedule = {flight: [slot for slot in
    ↪  maintenance_slots if y[flight, slot].solution_value() ==
    ↪  1] for flight in flights}
    return crew_assignment, maintenance_schedule
else:
    print('No optimal solution found.')
    return None, None

# Example data
```

```
crews = ['c1', 'c2', 'c3']
flights = ['f1', 'f2']
maintenance_slots = ['m1', 'm2']
flight_crew_requirements = {'f1': 2, 'f2': 1}
crew_availability = {'c1': {'f1': True, 'f2': True}, 'c2': {'f1':
↪    True, 'f2': False}, 'c3': {'f1': False, 'f2': True}}
maintenance_requirements = {'f1': 1, 'f2': 1}

crew_assignment, maintenance_schedule =
↪    crew_assignment_optimization(
    crews, flights, maintenance_slots, flight_crew_requirements,
    ↪    crew_availability, maintenance_requirements)

print("Crew Assignment:", crew_assignment)
print("Maintenance Schedule:", maintenance_schedule)
```

This code defines the necessary components for optimizing crew assignments and maintenance schedules:

- `crew_assignment_optimization` function formulates the optimization constraints and objective using linear programming.

- Decision variables `x` and `y` represent crew and maintenance assignments, respectively.

- Constraints ensure flights are staffed with required crew and meet maintenance needs while respecting crew availability.

- The objective is to maximize the number of flights with optimal crew and maintenance allocations.

- Example data demonstrates typical inputs and outputs using SCIP solver for optimization.

The final block provides a demonstration of the code execution with a sample dataset.

Chapter 95

Computational Methods for Wind Turbine Aerodynamics

Below is a Python code snippet that encompasses the core computational elements for simulating and optimizing the aerodynamic performance of wind turbines using the Blade Element Momentum (BEM) theory. This implementation includes the calculation of rotor blade forces, power output estimation, and optimization procedures for maximizing efficiency based on aerodynamic analyses.

```python
import numpy as np

def bem_analysis(rotor_radius, num_blades, wind_speed, air_density,
 ↪  tip_speed_ratio, pitch_angle):
    '''
    Perform Blade Element Momentum (BEM) analysis for a wind
     ↪  turbine.
    :param rotor_radius: Radius of the rotor.
    :param num_blades: Number of blades.
    :param wind_speed: Wind speed in m/s.
    :param air_density: Air density in kg/m^3.
    :param tip_speed_ratio: Tip speed ratio of the rotor.
    :param pitch_angle: Blade pitch angle in degrees.
    :return: Power output and efficiency of the turbine.
    '''
    # Discretize the blade into elements
    num_elements = 20
    dr = rotor_radius / num_elements
```

```python
    r = np.linspace(dr / 2, rotor_radius - dr / 2, num_elements)  #
    ↪  Radial positions

    # Define airfoil characteristics (simplified for illustration)
    Cl = 1.0  # Lift coefficient, may vary with angle of attack
    Cd = 0.01  # Drag coefficient, may vary with angle of attack

    # Initialize power and thrust
    total_power = 0
    total_thrust = 0

    # Calculate the angular velocity of the rotor
    rotor_angular_velocity = tip_speed_ratio * wind_speed /
    ↪  rotor_radius

    for ri in r:
        # Calculate local flow velocity
        local_speed = np.sqrt((rotor_angular_velocity * ri) ** 2 +
        ↪  wind_speed ** 2)
        phi = np.arctan2(wind_speed, rotor_angular_velocity * ri) *
        ↪  180 / np.pi - pitch_angle  # Flow angle of attack
        # Convert pitch angle to radians
        pitch_rad = np.radians(pitch_angle)

        # Calculate lift and drag per unit length
        lift = 0.5 * air_density * local_speed ** 2 * Cl * dr
        drag = 0.5 * air_density * local_speed ** 2 * Cd * dr

        # Resolve forces into the rotor plane
        thrust = lift * np.cos(pitch_rad) + drag * np.sin(pitch_rad)
        torque = lift * np.sin(pitch_rad) * ri - drag *
        ↪  np.cos(pitch_rad) * ri

        # Sum the elemental forces
        total_thrust += thrust
        total_power += torque * rotor_angular_velocity

    # Calculate turbine efficiency
    # Maximum theoretical efficiency (Betz limit) is 59.3%
    max_efficiency = 0.593
    turbine_efficiency = total_power / (0.5 * air_density * np.pi *
    ↪  rotor_radius ** 2 * wind_speed ** 3)

    return total_power, turbine_efficiency * max_efficiency

# Example of BEM analysis
rotor_radius = 40  # in meters
num_blades = 3
wind_speed = 12  # m/s
air_density = 1.225  # kg/m^3 (standard sea level pressure)
tip_speed_ratio = 7
pitch_angle = 3  # degrees
```

```
power, efficiency = bem_analysis(rotor_radius, num_blades,
↪  wind_speed, air_density,
                              tip_speed_ratio, pitch_angle)

print("Power Output:", power, "Watts")
print("Turbine Efficiency:", efficiency * 100, "%")
```

This code defines several key functions necessary for simulating the aerodynamic performance of wind turbines:

- bem_analysis function performs the core simulation using the Blade Element Momentum theory. It calculates forces and power output based on various aerodynamic parameters of a wind turbine rotor.

- Parameters such as rotor_radius, num_blades, wind_speed, air_density, tip_speed_ratio, and pitch_angle define the operating and structural conditions for the simulation.

- The results include the computation of the power output and an estimation of turbine efficiency, considering fundamental aerodynamic interactions.

The example at the end of the snippet demonstrates how to use the defined functions to evaluate the power and efficiency of a specific wind turbine configuration.

Chapter 96

Algorithms for Supersonic Boom Minimization

Below is a Python code snippet that encompasses the core computational elements for designing aircraft shapes that minimize sonic boom intensity using computational fluid dynamics and optimization techniques.

```python
import numpy as np
from scipy.optimize import minimize
from scipy.integrate import solve_ivp

def objective_function(shape_params, *args):
    '''
    Objective function to minimize sonic boom intensity.
    :param shape_params: Parameters defining the aircraft shape.
    :return: Calculated sonic boom intensity.
    '''
    # Example placeholder for aerodynamic analysis
    intensity = np.sum(shape_params**2)  # Dummy calculation
    return intensity

def generate_initial_shape():
    '''
    Generate initial aircraft shape parameters.
    :return: Initial shape parameter array.
    '''
    return np.random.rand(10)  # Example shape parameter vector

def constraint_function(shape_params):
```

```
    '''
    Constraints on the shape parameters.
    :param shape_params: Aircraft shape parameters.
    :return: Constraint satisfaction (value should be 0).
    '''
    return shape_params - 0.1  # Dummy constraint to keep parameters
    ↪  positive

def sonic_boom_simulation(t, y, shape_params):
    '''
    Simulates the propagation of sound waves for given shape
    ↪  parameters.
    :param t: Time variable for integration.
    :param y: State vector of the wave.
    :param shape_params: Shape parameters of the aircraft.
    :return: Derivatives of the state vector.
    '''
    return -0.01 * y   # Simplified dynamics for illustration

def perform_sonic_boom_analysis(shape_params):
    '''
    Perform analysis to evaluate sonic boom characteristics.
    :param shape_params: Parameters defining the aircraft shape.
    :return: Analysis result as a float value.
    '''
    t_span = [0, 100]  # Dummy time range
    y0 = [1.0]   # Initial condition for sound wave
    sol = solve_ivp(sonic_boom_simulation, t_span, y0,
    ↪  args=(shape_params,))

    # Analysis result could be max/min of solution, or final value
    return np.max(sol.y[0])

# Constraint definition for optimization problem
constraints = [{'type': 'ineq', 'fun': constraint_function}]

# Initial shape parameters for the optimization
initial_shape = generate_initial_shape()

# Optimize the aircraft shape parameters to minimize sonic boom
↪  intensity
result = minimize(objective_function, initial_shape,
↪  constraints=constraints)

if result.success:
    optimized_shape = result.x
    analysis_result = perform_sonic_boom_analysis(optimized_shape)
    print("Optimized shape parameters:", optimized_shape)
    print("Sonic boom analysis result:", analysis_result)
else:
    print("Optimization did not succeed. Status:", result.message)
```

This code defines several key functions necessary for minimizing sonic boom intensity in aircraft design:

- `objective_function` assesses the sonic boom intensity based on input shape parameters.

- `generate_initial_shape` initializes a random set of shape parameters as starting points for optimization.

- `constraint_function` ensures shape parameters satisfy design constraints during optimization.

- `sonic_boom_simulation` simulates how sound waves propagate influenced by the aircraft's shape.

- `perform_sonic_boom_analysis` evaluates the final sonic boom impact due to optimized shape parameters.

The code optimizes aircraft design for minimal sonic booms using the defined functions and constraint methods to demonstrate a simplified model.

Chapter 97

Computational Modeling of Environmental Control Systems

Below is a Python code snippet that encompasses the core computational elements of simulating systems that regulate cabin pressure, temperature, and air quality within aerospace systems.

```python
import numpy as np

class CabinEnvironmentSimulator:
    def __init__(self, initial_pressure, initial_temperature,
    ↪    initial_air_quality):
        '''
        Initialize the simulator with initial conditions.
        :param initial_pressure: Initial pressure in the cabin in
        ↪    Pascals.
        :param initial_temperature: Initial temperature in Celsius.
        :param initial_air_quality: Initial air quality index.
        '''
        self.pressure = initial_pressure
        self.temperature = initial_temperature
        self.air_quality = initial_air_quality

    def update_pressure(self, external_pressure, leak_factor):
        '''
        Update the cabin pressure based on external pressure and
        ↪    leak factor.
```

```python
        :param external_pressure: External pressure in Pascals.
        :param leak_factor: The factor of leakage, range [0,1].
        '''
        pressure_diff = external_pressure - self.pressure
        self.pressure += leak_factor * pressure_diff

    def update_temperature(self, target_temperature,
    ↪   control_efficiency):
        '''
        Update the cabin temperature towards a target temperature.
        :param target_temperature: Desired cabin temperature in
        ↪   Celsius.
        :param control_efficiency: Efficiency of the temperature
        ↪   control system [0,1].
        '''
        temperature_diff = target_temperature - self.temperature
        self.temperature += control_efficiency * temperature_diff

    def update_air_quality(self, air_filters_efficiency,
    ↪   contamination_sources):
        '''
        Update the cabin air quality index.
        :param air_filters_efficiency: The efficiency of the air
        ↪   filters [0,1].
        :param contamination_sources: The number of contamination
        ↪   sources in the cabin.
        '''
        self.air_quality -= air_filters_efficiency *
        ↪   contamination_sources
        if self.air_quality < 0:
            self.air_quality = 0

    def simulate(self, steps, external_conditions):
        '''
        Simulate the cabin environment over a number of time steps.
        :param steps: Number of simulation steps.
        :param external_conditions: List of tuples
        ↪   (external_pressure, contamination_sources) for each
        ↪   step.
        '''
        for step in range(steps):
            external_pressure, contamination_sources =
            ↪   external_conditions[step]
            self.update_pressure(external_pressure,
            ↪   leak_factor=0.05)
            self.update_temperature(target_temperature=25,
            ↪   control_efficiency=0.8)
            self.update_air_quality(air_filters_efficiency=0.9,
            ↪   contamination_sources=contamination_sources)
            print(f"Step {step}: Pressure={self.pressure} Pa,
            ↪   Temperature={self.temperature} C, Air
            ↪   Quality={self.air_quality}")
```

```
# Simulating cabin conditions for 10 time steps with varying
↪   external pressures and contamination sources
initial_conditions = (101325, 20, 100)
environment_simulator =
↪   CabinEnvironmentSimulator(*initial_conditions)
external_conditions = [(101000 + i*100, i) for i in range(10)]

environment_simulator.simulate(10, external_conditions)
```

This code defines several key functions necessary for simulating the cabin environment:

- The CabinEnvironmentSimulator class initializes with initial cabin pressure, temperature, and air quality.

- update_pressure function that adjusts cabin pressure based on external pressure and a leak factor.

- update_temperature function for moving cabin temperature towards a target through a control efficiency parameter.

- update_air_quality function that reduces the air quality index accounting for air filter efficiency and contaminants.

- simulate is used to iterate over several steps, applying updates at each step based on external conditions.

The final block of code demonstrates simulating cabin conditions over 10 steps with varying external factors.

Chapter 98

Algorithms for Space Mission Analysis and Design

Below is a Python code snippet that encompasses the core computational elements of aerospace mission planning, including payload optimization and launch window analysis using a simplified heuristic method for demonstration purposes.

```python
import numpy as np
from scipy.optimize import minimize

def payload_optimization_function(mass_list, capacity, profit_list):
    '''
    Optimizes the selection of payloads to maximize profit within a
    ↪  capacity constraint.
    :param mass_list: List of masses for each payload.
    :param capacity: Total capacity available for the payloads.
    :param profit_list: Corresponding list of profits for each
    ↪  payload.
    :return: A list indicating which payloads to take.
    '''

    num_payloads = len(mass_list)
    # Objective function: negative sum of profits (since we're
    ↪  maximizing)
    def objective(x):
        return -np.dot(profit_list, x)

    # Constraints: sum of masses must be less than or equal to
    ↪  capacity
```

```python
    constraints = ({'type': 'ineq', 'fun': lambda x: capacity -
    ↪   np.dot(mass_list, x)})

    # Bounds: all x should be between 0 and 1 (either take or don't
    ↪   take each payload)
    bounds = [(0, 1) for _ in range(num_payloads)]

    # Initial guess (no payloads taken)
    x0 = np.zeros(num_payloads)

    # Optimize
    result = minimize(objective, x0, bounds=bounds,
    ↪   constraints=constraints, method='SLSQP')

    return result.x

def launch_window_analysis(launch_dates, success_probabilities):
    '''
    Analyzes a set of potential launch dates to maximize mission
    ↪   success probability.
    :param launch_dates: List of candidate launch dates.
    :param success_probabilities: Corresponding success
    ↪   probabilities for each launch date.
    :return: Launch date with the highest success probability.
    '''
    # Select the launch date with the highest probability
    best_date_index = np.argmax(success_probabilities)
    return launch_dates[best_date_index]

# Example payload data
masses = [300, 400, 600, 500]  # masses in kg
capacity_limit = 1000  # capacity limit in kg
profits = [1000, 1200, 1500, 1600]  # profits in arbitrary units

# Example launch window data
launch_dates = ['2023-11-01', '2023-12-15', '2024-01-20']
success_probabilities = [0.85, 0.9, 0.8]

# Payload optimization
optimal_payloads = payload_optimization_function(masses,
↪   capacity_limit, profits)
print("Optimal Payload Selection:", optimal_payloads)

# Launch window analysis
best_launch_date = launch_window_analysis(launch_dates,
↪   success_probabilities)
print("Best Launch Date:", best_launch_date)
```

This code snippet defines essential functions necessary for aerospace mission planning:

- **payload_optimization_function** selects the best combination of payloads to maximize profit whilst adhering to capacity constraints using linear programming.

- `launch_window_analysis` determines the most optimal launch date based on maximizing the probability of mission success.

The example payload data includes masses and profits, showcasing how the optimization process chooses the payloads. The launch window example evaluates options to identify the most favorable launch date.

Chapter 99

High-Performance Computing in Aerospace Simulations

Below is a Python code snippet that encompasses the implementation of algorithms optimized for supercomputing environments to tackle large-scale aerospace problems, focusing on parallel processing using the mpi4py library.

```python
from mpi4py import MPI
import numpy as np

# Initialize the MPI communicator
comm = MPI.COMM_WORLD
rank = comm.Get_rank()
size = comm.Get_size()

def compute_aero_forces(velocity_field, pressure_field):
    '''
    Simulate the computation of aerodynamic forces.
    :param velocity_field: The velocity field array.
    :param pressure_field: The pressure field array.
    :return: Simulated aerodynamic force value.
    '''
    # Simplified dummy computation
    force_x = np.sum(velocity_field) * 0.1
    force_y = np.sum(pressure_field) * 0.2
    return force_x, force_y

def distributed_computation(n_elements):
    '''
```

```python
    Distribute elements across processes and perform computation.
    :param n_elements: Total number of elements to process.
    :return: Computed forces from all processes.
    '''
    # Scatter the workload evenly among processes
    elements_per_proc = n_elements // size
    elements = np.arange(rank * elements_per_proc, (rank + 1) *
    ↪   elements_per_proc)

    # Initialize dummy velocity and pressure fields
    velocity_field = np.random.rand(elements_per_proc)
    pressure_field = np.random.rand(elements_per_proc)

    # Compute aerodynamic forces
    local_forces = compute_aero_forces(velocity_field,
    ↪   pressure_field)

    # Gather results
    global_forces = comm.gather(local_forces, root=0)

    return global_forces

def main():
    # Total number of aerodynamic elements
    n_elements = 10000  # Example size, can be scaled

    # Perform distributed computation
    forces = distributed_computation(n_elements)

    if rank == 0:
        # Aggregate global forces
        total_force_x = sum(f[0] for f in forces)
        total_force_y = sum(f[1] for f in forces)

        print(f"Total Aerodynamic Force (X): {total_force_x}")
        print(f"Total Aerodynamic Force (Y): {total_force_y}")

if __name__ == "__main__":
    main()
```

This snippet demonstrates the use of MPI through mpi4py for distributing computational tasks in aerodynamic simulations, suitable for large-scale aerospace problems:

- compute_aero_forces function simulates the calculation of aerodynamic forces using dummy velocity and pressure fields.

- distributed_computation function divides the computational workload among available processes and gathers computed results across MPI processes.

- The main execution begins with initializing total elements to be processed, invoking distributed computations, and aggregating results.

- Processes communicate using MPI, effectively optimizing computations for supercomputing environments.

This approach enhances computational efficiency, making it feasible to tackle complex aerodynamic simulations across multiple processors.

www.ingramcontent.com/pod-product-compliance
Lightning Source LLC
Chambersburg PA
CBHW071446220526
45472CB00003B/685